JN044891

Understanding Interfacial Reaction Kinetics in the Fe-C-O system

Yasushi SASAKI

Tohoku University Press, Sendai

Published in 2022 by
Tohoku University Press, Sendai
2-1-1 Katahira, Aoba-ku, Sendai 980-8577, Japan
Tel : +81-22-214-2777
Fax : +81-22-214-2778
Website : https://www.tups.jp/
E-mail : info@tups.jp

ISBN978-4-86163-376-8

Printed in Japan 2022 by
TOHOKU UNIVERSITY CO-OP

To
Akiko SASAKI

Copyright and figure acknowledgements

Preface

There are many books on chemical reaction kinetics. Most of the kinetics books are focussed on homogeneous reactions, but few are focused on heterogeneous reactions despite their importance. The characteristics of heterogeneous reaction kinetics are very different from those of homogeneous reactions. Thus, the analytical method used in homogeneous reactions kinetics cannot be directly applied to heterogeneous reactions, and the study heterogeneous reaction kinetics requires special analytical tools.

The author intends to provide a textbook that will serve as an introductory course in heterogeneous reaction kinetics for metallurgy, materials science, chemical engineering, and associated discipline students. The purpose of this book is to present a precise, concise, and relatively complete treatment of heterogeneous reaction kinetics. Throughout, the primary aim is to clarify the underlying physical meaning and implications of the concepts of heterogeneous reaction kinetics. Kinetics is strongly related to thermodynamics. The knowledge of thermodynamics is critical to understand kinetics. The primary focus of this book is the relationship between kinetics and thermodynamics.

Consistent with the emphasis on brevity and fundamentals, experimental results are quoted as often as possible to help the reader understand subjects. The chapters are mostly independent of one another, and the reader can access the topic of their interest without necessarily having to read the preceding sections.

Heterogeneous reactions are classified into two fields: catalytic and non-catalytic reactions. Catalytic reactions strongly depend on the structure of catalysts, and the range of discussion in this book does not extend to these reactions. This book mainly focuses on the kinetics of non-catalytic reactions, especially the metallurgical reaction of the Fe-C-O system, as indicated in the title. By limiting the subject of interest to the area of the Fe-C-O system, the author hopes that the reader can establish a unified and integrated idea about kinetics and gain a clearer understanding of kinetics instead of merely accumulating many fragmented pieces of knowledge.

Metallurgical reactions of the Fe-C-O system contain many important fundamental reactions, such as the iron oxide reduction, iron oxidation, and iron carburization in iron/steelmaking processes. A considerable amount of research work has been carried out for an understanding of these reactions not only from the theoretical perspective but also from their industrial importance. These reactions provide many good examples to demonstrate the principles of kinetics.

First of all, I would like to thank the late Prof. G. R. Belton, who led me to the wonderful world of reaction kinetics. I also wish to thank Prof. O. Ostrovsky (UNSW Australia) for being excellent host during a three years stay, which made it possible for a great deal of progress to be made on the text. Some content of this book was used over three years for a kinetics course taught to students at POSTECH (South Korea). The experience gained from those colleagues and students (Hyun-Soo, Sun-Young, In-Hyeon, M. Ji-Won, Woo-II, J. Ji-Won, Ji-Ook, and Weon-Hui) guided the author in putting together into the book. Dr. Lu Xin at Tohoku University read manuscript carefully and made perspective comments that I did my best to incorporate. I wish to express my gratitude to him.

Thanks to all the authors and the publishers of the many papers I have used, in some cases adapting diagrams. Finally, to my wife, Akiko, who has received much less of my time and attention than is her due.

Contents

4

Fundamentals

Chapter 1
Introduction

Chapter 1   Introduction

1.1   What is reaction kinetics?

There are two main reasons to study reaction kinetics; (1) to control the chemical process and (2) to clarify the reaction mechanism.

With regard to reason (1), it is widely assumed that reaction kinetics concerns determining the reaction rates and establishing their rate constants. Indeed, information of the reaction constant is essential in the design of reactors. However, the most important aim of reaction kinetics is to elucidate the reaction mechanism of the reaction. Despite what many think, finding the reaction rate equations, or evaluating reaction constants of the reactions, is not the final aim of reaction kinetics. These are the means to finding the reaction mechanism. Then what is the reaction mechanism in the field of reaction kinetics?

Chemical reactions rarely proceed in one process. The conventional way of writing the stoichiometric equation of a chemical reaction, such as $2CO + O_2 \rightarrow 2CO_2$, indicates the initial and final states of the reaction system. The vast majority of reactions occur in several reaction steps that contain intermediate species which do not appear in the stoichiometric reaction equations. This series of reactions is referred to as a "reaction mechanism" in chemical kinetics. Each step that contributes to the overall reaction is termed an "elementary step" or "elementary reaction".

The reaction between hydrogen and halogens can be an excellent example to understand the reaction mechanism. Bodenstein (1894) carried out a series of kinetics studies of the reaction between hydrogen and halogens. The following chemical equation expresses the reaction between hydrogen and iodine.

$$H_2 + I_2 \rightarrow 2HI \qquad\qquad \text{----- (1-1)}$$

He found that the forward reaction rate was expressed by

$$d[HI]/dt = k\,[H_2][I_2] \qquad\qquad \text{----- (1-2)}$$

where $[H_2]$ and $[I_2]$ are the concentration of H_2 and I_2, respectively, and k is the rate constant. In this book, the concentration of component X is denoted by [X]. From eq. (1-2), it seems that the reaction (1-1) simply proceeds with the collision between H_2 and I_2 molecules, as written in the equation (1-2). That is, this reaction seems to be an elementary reaction. Then Bodenstein and S.C.Lind (1907) measured the rates of the reaction between hydrogen and bromine

$$H_2 + Br_2 \rightarrow 2HBr \qquad\qquad \text{----- (1-3)},$$

They expected a similar rate equation with the reaction (1-1); however, the obtained rate equation was more complex than anticipated. The overall reaction rate equation was expressed by,

$$\frac{d[HBr]}{dt} = \frac{2k_2\left(\dfrac{k_3}{k_4}\right)\left(\dfrac{k_1}{k_5}\right)^{\frac{1}{2}}[H_2][Br_2]^{\frac{1}{2}}}{k_3/k_4 + [HBr]/[Br_2]} \qquad\qquad \text{----- (1-4).}$$

where k_1 to k_5 are parameters determined by experiments. The rate equation (1-4) strongly suggested that the reaction mechanism of (1-3) is more complicated. They could not explain why the equation (1-3) expressed the reaction rate (1-4). Later, Herzfeld, Christiansen, and Polyanyi independently and successfully explained the reaction rate dependency (1-4) by assuming the following reaction steps.

$$Br_2 \rightarrow Br + Br \qquad\qquad ----- (1\text{-}5)$$
$$Br + H_2 \rightarrow HBr + H \qquad\qquad ----- (1\text{-}6)$$
$$H + Br_2 \rightarrow HBr + Br \qquad\qquad ----- (1\text{-}7)$$
$$Br + Br \rightarrow Br_2 \qquad\qquad ----- (1\text{-}8)$$

The reactions (1-6) and (1-7) form a 'chain' (the Br radical consumed in reaction (1-6) is regenerated in reaction (1-7)), and this occurs many times before the sequence of the reaction series is terminated by reaction (1-8). In reaction kinetics, the set of the elementary reaction (1-5) to (1-8) is called the reaction mechanism of the reaction (1-3). Details of the deduction of the rate equation (1-4) are shown in appendix A.

The precise relationship between the reaction rate and the concentration of reactants is purely empirical and cannot be determined from the stoichiometry of reactions. Namely, chemical kinetics is purely experimental science, much like thermodynamics. Finding out elementary reactions is not straightforward since several mechanisms tend to be consistent with an observed overall rate law. We must guess and then verify them with experiments until confirmed by experimental data.

Chemical kinetics involves a series of elementary reactions rather than each elementary reaction itself. An elementary step represents the reaction at the molecule level and is treated by molecular dynamics in quantum chemistry. Molecular dynamics handle the elementary reaction as a recombinant process of chemical bonds due to the collision of atoms and molecules. In contrast, chemical kinetics is a macroscopic science and deals with the average behavior of reacting substance changes involving the molar order: at least 6×10^{23} molecules or more. This book focuses on chemical kinetics but not on molecular dynamics.

Stoichiometric equations, such as equation (1-1), are generally called reaction equations. As already explained, the stoichiometric reaction equation simply expresses the mass balance and does not show the reaction mechanism except in elementary reactions.

1.2   The relation between reaction kinetics and thermodynamics
Law of mass action

Thermodynamics, like reaction kinetics, is a useful and powerful tool in the study of chemistry. Thermodynamics deals with chemical systems at equilibrium, and their properties do not change with time. Most real reaction systems are not at equilibrium, and chemical reactions occur to seek equilibrium. Reaction kinetics focus on these changes in chemical properties in time. Because time is not considered in thermodynamics, it cannot be used to evaluate reaction rates. However, reaction kinetics have one contact point with thermodynamics through the *law of mass action* for reversible reactions.

A reversible reaction consists of two reaction steps in the opposite direction. In a simple reversible reaction of A + B = C + D, substances A and B react to form C and D. In a reverse reaction, C and D react to form A and B. Reversible reactions will reach an equilibrium state where the concentrations of the reactants and products no longer change.

For a simple reversible reaction between reactants A and B to give products C and D, the apparent reaction can be expressed by

$$aA + bB = cC + dD \qquad\qquad ----- (1\text{-}9)$$

At equilibrium, the following relation among the concentrations of reactants and products in the reaction system (1-9) is well known,

$$K = \frac{[C_E]^c \, [D_E]^d}{[A_E]^a \, [B_E]^b} \qquad\qquad ----- (1\text{-}10)$$

where $[A_E]$, $[B_E]$, $[C_E]$, $[D_E]$ are the equilibrium concentrations of A, B, C, and D, respectively. K is called the equilibrium constant. **The relation (1-10) is called the law of mass action and is one of the most important thermodynamic relations**.

Now, the law of mass action is thermodynamically derived using the chemical potentials of each substance in a reaction system. Historically, this law was obtained by Goldberg and Waage (1864) based on the reaction kinetics perspective [1]. Based on two assumptions, they deduced the mass action law.

The assumption I: The rates of a chemical reaction are directly proportional to the product of the molar concentrations of reacting substances, with each concentration term raised to the power. The power values are assumed to be equal to the stoichiometric coefficients of the substances in the chemical equation. Then, the forward and backward reaction rates in the reversible reaction (1-9) can be expressed by

$$\text{Forward reaction rate} = k_f \, [A]^a \, [B]^b \qquad ----- (1\text{-}11)$$
$$\text{Backward reaction rate} = k_b \, [C]^c \, [D]^d \qquad ----- (1\text{-}12)$$

k_f and k_b are proportional constants and are called a forward and backward reaction rate coefficient, respectively. It was also assumed that the rate equation (1-11) and (1-12) could be applicable even under the equilibrium condition.

Assumption II: The forward reaction rate is equal to the backward reaction rate at equilibrium. This assumption is a great insight. It gives us the dynamic view of the equilibrium state instead of the static view used before it.

Based on the assumption I and II, and substituting each equilibrium concentration $[A_E]$, $[B_E]$, $[C_E]$, and $[D_E]$ into the equation (1-11) and (1-12) yields,

$$\text{Forward reaction rate at equilibrium condition} = k_f \, [A_E]^a \, [B_E]^b \qquad ….. (1\text{-}11)'$$
$$\text{Backward reaction rate at equilibrium condition} = k_b \, [C_E]^c \, [D_E]^d \qquad ….. (1\text{-}12)'$$

Equating the rate of forwarding reaction with the rate of backward reaction yields,

$$k_f [A_E]^a [B_E]^b = k_b [C_E]^c [D_E]^d \qquad \text{----- (1-13)}$$

Since k_f and k_b are constant, the ratio k_f/k_b is also constant and is represented by Kc.

$$\text{Kc} = k_f/k_b$$
$$= [C_E]^c [D_E]^d/[A_E]^a [B_E]^b \qquad \text{----- (1-14)}$$

The equation (1-14) is essentially the same as the equation (1-10). Namely, based on the re-action kinetics, the law of mass action is deduced. The constant Kc is called an equilibrium constant. Although the law of mass action is deduced based on the reaction equations, it does not necessarily ensure that the proposed reaction mechanism is correct. A typical error is writing down the rate equations based on the stoichiometric reaction equations since the mass action law was obtained using these reaction equations.

Then, why was the mass action law deduced by using the reaction equations?

When assuming *any* possible combinations of elementary reactions for the reversible reaction, we can deduce the correct mass action law as long as ***all these elementary reactions are at equilibrium***.

Pimentel and Spratley discussed the details of this popular error in their famous thermodynamics textbook [2]. They assumed two different mechanisms for the reaction between H_2 and I_2 (reaction (1-1)). One is the elementary reaction mechanism, where H_2 and I_2 molecules collide and produce two HI molecules. Another is a chain reaction mechanism that includes intermediates I and H atoms. The correct law of mass action is deduced if the assumed reaction mechanism satisfies the reaction equation! That is, **the derivation of the mass law based on reaction kinetics has nothing to do with the actual reaction mechanism**.

1.3   Why does a reaction occur?

Thermodynamics cannot provide us with the ***reaction rates***, but it can provide information on whether reactions will occur or not. This is possible with the use of the Gibbs free energy changes (ΔG) of the reactions. When the value of ΔG is positive, the reaction does not occur. When it has a negative value, the reaction does occur.

Why do the negative values of ΔG indicate the reaction will go ahead? The driving force for a reaction is not moving to lower-energy (or, more precisely, lower-enthalpy) products. Rather the driving force is concerned with entropy and is governed by the 2nd law of thermodynamics. The 2nd law allows us to predict which reaction will go ahead and which will not. The 2nd law says that any process which takes place is always associated with an increase in the ***entropy of the universe***.

Generally, the entropy change of the universe during a reaction is expressed by

$$\Delta S_{uni} = \Delta S_{surr} + \Delta S_{sys} \qquad \text{----- (1-15)}$$

where ΔS_{uni}, ΔS_{surr}, and ΔS_{sys} are entropy changes of the universe, the surroundings, and the system. ΔS_{sys} can be calculated from the lists in thermodynamics tables, but it seems difficult to evaluate the ΔS_{surr}. However, thermodynamics tells us that the change in the entropy of the universe is correlated with the Gibbs free energy change (ΔG), as represented by the following equation,

$$\Delta G = -\Delta S_{uni} / T \qquad\qquad ----- (1\text{-}16)$$

where T is the absolute temperature. The details of the deduction of Eq. (1-16) are presented in appendix B. From equation (1-16), ΔG has a negative value when the entropy of the universe increases. Thus, we can say that the reaction will occur when the value of ΔG is negative since T is always positive. **The sign of ΔG (the Gibbs free energy of the system) but not of ΔG⁰ (the change in the Standard Gibbs free energy) indicates whether the reaction will occur or not**. The change in *ΔG* in an endothermic reaction is schematically shown in Fig. 1-1. In an endothermic reaction, A → B, the product enthalpy H_B has a larger value than that of H_A when TS_B has a much larger positive value than TS_A, of the reactant. As a result, the Gibbs free energy change *ΔG* becomes negative, as shown in Fig. 1.1, and the reaction will occur even when $H < 0$. Instead of entropy S, TS is used to represent the same unit as Gibbs free energy G and enthalpy H.

It is noted that the reaction rates are not related to the magnitude of *ΔG* at all. The change in the Gibbs free energy only provides information about the possibility of the reaction's occurrence but not that of the reaction rate.

The diagram shown in Fig. 1-2 is called a reaction coordinate diagram. It shows how the energy of the system changes during a chemical reaction of A → B. The vertical axis represents the overall energy for substances, while the horizontal axis is the reaction coordinate. Tracing from left to right takes us through the reaction from the starting reactants to the final products, with practically the same reaction time.

As shown in Fig. 1-2, the reacting system passes through an energetically high and unstable state during the reaction from the reactant to the product. This unstable state is called the

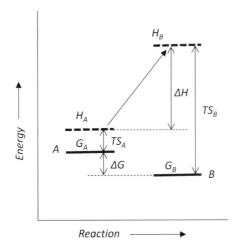

Fig. 1-1 The change in the Gibbs free energy, *ΔG*, in an endothermic reaction.

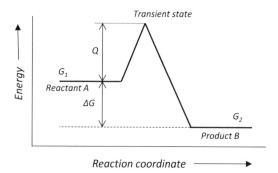

Fig. 1-2 Reaction profile for reaction A → B. G1 and G2 are the Gibbs free energies of reactant A and product B, respectively. Q is the activation energy for the reaction A → B.

"transition state" and can be calculated in quantum chemistry. The transient state has higher energy than the reactants. It is called the activation energy Q and is a measure of how difficult it is for the reaction to go ahead in any reasonable amount of time. This energy Q is required for the rearrangement of their valence electrons to form the products. The reactions have to overcome this energy barrier which corresponds to the activation energy (Q) to produce products. Namely, the reaction rate is determined by the *activation energy Q* but not the Gibbs free energy difference $\Delta G\ (= G_2 - G_1)$ between these of the reactants (G_1) and products (G_2).

1.4 Characteristics of heterogeneous reactions

In the previous sections, the general concept of reaction kinetics is introduced. As including "interfacial reactions" in the title of this book implies, the book is mainly focused on heterogeneous reaction kinetics, which have characteristics quite different from those of homogeneous reactions. In this section, the unique characteristics of the heterogeneous reaction are briefly outlined. Heterogeneous reactions take place at the surface of the solids or liquids. The surface involved in the reaction gives rise to several concerns in the study of heterogeneous reaction kinetics, (1) the effect of surface structure, and (2) the evaluation of the reaction area.

1.4.1 Effect of surface structure on reaction rates

As one would expect, the heterogeneous reaction rates are affected by the surface structures. In other words, their rates depend on not only the reactants but also on the surface structures. A naive view of the surface of solid materials is that it is simply a flat layer of atoms, but detailed studies have demonstrated that surfaces themselves have complex structures. An example is the structure of the surface of wüstite. As shown in Fig. 1-3, the surface structure of wüstite has several surface orientations and has a fairly complicated structure, complete with surface defects. The surface defects on a solid surface are schematically shown in Fig. 1-4. Most atoms are arranged in terraces, and terraces often have surface defects, such

(649)

(10 1 11)

(7 3 12)

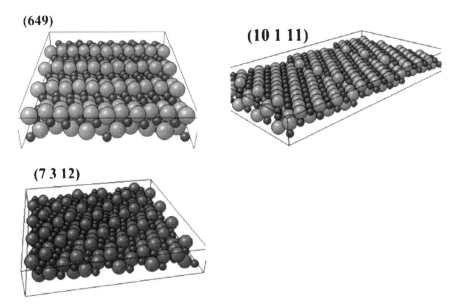

Fig. 1-3 The surface structures of wüstite with several surface orientations. The large and small spheres correspond to oxygen and Fe.

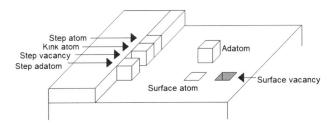

Fig. 1-4 Schematic drawing of solid surface with various surface defects.

as adatoms (adsorbed atoms) or terrace vacancies. The areas between terraces are called steps. These steps are not necessarily straight, but rather have kinks and step-adatoms.

The surface structure is of critical importance in reaction kinetics since the reaction rate may vary by several orders of magnitude, depending on the number of steps or the arrangement of adatoms since the reaction initiates from these structural defects. That is, the reaction rates at the surface can be significantly influenced by the surface's structure. Bahagat et al. [3] observed the partially reduced wüstite surface with grains with (4 1 10) and (7 2 10) orientations, as shown in Fig. 1-5. The white spots at the grey-colored wüstite surface are the produced small Fe crystals. Many Fe crystals are observed at the (4 1 10) plane but fewer can be seen at the (7 2 10) plane. This demonstrates how strongly the reduction rates depend on the grain orientations.

Unlike solid surfaces, liquid surfaces are *homogeneously* random or have no fixed struc-

16

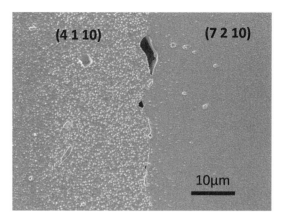

Fig. 1-5 Partially reduced wüstite surface with grains with (4 1 10) and (7 2 10) orientations.

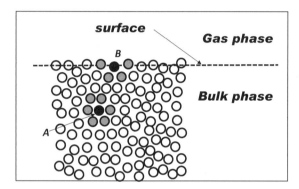

Fig. 1-6 Schematic atomic configurations in bulk and surface.

tures since the surface molecules constantly move, and any non-uniformity is short-lived. The atomic configurations in the bulk and at the surface are schematically shown in Fig. 1-6.

As shown in Fig. 1-6, atom A presented by the closed circle in the bulk phase is surrounded by 6 atoms (in grey), while atom B at the surface binds with 4 atoms (in grey). There are fewer chemical bindings at the surface of the upper side than those in the bulk, so the relative energies of the surface atoms are higher than those in the bulk atoms. Therefore, atoms at the surface seek other surface atoms to bind to so that they can stabilize. It means that the number of surface atoms must be maximized to facilitate the binding of as many surface atoms as possible. As a result, the time-averaged liquid surface possibly resembles a 2-dimensionally closed packing structure, or (111) plane of FCC crystals, although these atoms are more loosely packed than those at solid-state as shown in Fig. 1-7.

Each atom at the liquid surface moves incessantly and does not stay at the fixed position. However, the time-averaged positions of atoms at the surface correspond to the layout of the loosely packed (111) plane of FCC crystals.

In this aspect, the liquid surface has essentially only one unique structure, while the surface

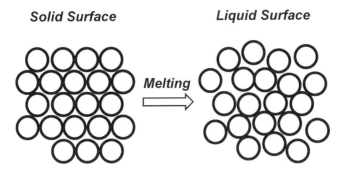

Solid Surface   **Liquid Surface**

Melting

Fig. 1-7 View from above the surface of the schematic atom configuration for the solid and liquid surface.

structures of polycrystalline materials have almost infinite variations. Thus, the reaction rate of the liquid surface has intrinsic value, and this depends only on the concerned liquid once the reaction condition is fixed. However, the actual situation is not so simple. We must think about the effect of impurities, or the "adsorption process."

1.4.2 Evaluation of reaction area.

Even much-purified materials always contain a small number of impurities. The atoms at the surface bind together to minimize the energy. However, their energies are still higher than those inside. Thus, the surface atoms have a preference to bind with these impurities to decrease their energies. Some impurities, such as O and S, strongly bind with the atoms at the surface, in effect enriching these impurity elements at the surface. These elements are called surface-active elements. This surface enrichment phenomenon is called adsorption or segregation (for solid cases). The adsorption phenomena inevitably occur on all surfaces of solids and liquids. Further details of the adsorption phenomena can be found in Chapter 5.

These surface-active elements are known to block the reaction sites and reduce the apparent reaction rates since the reaction only occurs at the unblocked bare surface. For example, the surface coverage of S atoms in liquid Fe-0.1 mass % S at 1873K has been shown to be about 70%! That is, only 30% of the surface is available for the reaction. Thus, to evaluate the heterogeneous reaction rate at the bare surface, we need to know the adsorption behavior of these adsorbed species to estimate the reacting area.

Further reading
K. J. Laidler: Chemical kinetics, 3rd ed., Harper and Row, New York, 1987.

J. Keeler and P. Wothers: Why chemical reactions happen: Oxford Uni. Press. Oxford UK, 2003.

H. Wise and J. Oudar: Material concepts in surface reactivity and catalysis. Dover Pub. Inc., New York 1990.

18

References

[1] E. W. Lund: J. Chem. Edu., 42, 1965, 548-550.

[2] G. C. Pimentel and R. D. Spratley: Understanding chemical thermodynamics, Holden-Day Inc., 1969, San Francisco, (1969), 16-23.

[3] M. Bahgat: Private communication.

Appendix A

Reaction mechanism of the reaction between H_2 and Br_2

Hydrogen-bromine reaction is expressed by,

$$H_2 + Br_2 \rightarrow 2HBr \qquad \text{----- (A-1)}$$

The reaction (A-1) occurs by the following elementary reaction steps.

$$Br_2 \rightarrow Br + Br \qquad \text{----- (A-2)}$$
$$Br + H_2 \rightarrow HBr + H \qquad \text{----- (A-3)}$$
$$H + Br_2 \rightarrow HBr + Br \qquad \text{----- (A-4)}$$
$$Br + Br \rightarrow Br_2 \qquad \text{----- (A-5)}$$

Because the reactivity of atomic H and Br is very large, the production and consumption rates can be assumed to be the same, or produced H and Br are spontaneously consumed.

$$d[Br]/dt = 2k1[Br_2] - k2[Br][H_2] + k3[H][Br_2] + k4[H][HBr] - 2k5[Br_2]^2 = 0$$
$$\text{----- (A-6)}$$

$$d[H]/dt = k2[Br][H_2] - k3[H][Br_2] - k4[H][HBr] = 0 \qquad \text{----- (A-7)}$$

From the above simultaneous equations,

$$[Br] = [(k1/k5)\,[Br_2]]^{1/2} \qquad \text{----- (A-8)}$$

$$[H] = \frac{k_2\left(\dfrac{k_1}{k_5}\right)^{\frac{1}{2}}[H_2][Br_2]^{\frac{1}{2}}}{k_3[Br_2] + k_4[HBr]} \qquad \text{----- (A-9)}$$

For the formation of HBr,

$$\frac{d[HB_r]}{dt} = k2[Br][H2] + k3[H][Br2] - k4[H][HBr] \qquad \text{----- (A-10)}$$

Substitute [Br] and [H] into the equation (A-10),

$$\frac{d[HBr]}{dt} = \frac{2k_2\left(\dfrac{k_3}{k_4}\right)\left(\dfrac{k_1}{k_5}\right)^{\frac{1}{2}}[H_2][Br_2]^{\frac{1}{2}}}{k_3/k_4 + [HBr]/[Br_2]} \qquad \text{----- (A-11)}$$

The equation (A-11) is the same as the experimentally obtained rate equation.

Appendix B Gibbs free energy

Generally, the entropy change of the universe is expressed by

$$\Delta S_{uni} = \Delta S_{surr} + \Delta S_{sys} \qquad \text{----- (B-1)}$$

where ΔS_{uni}, ΔS_{surr}, and ΔS_{sys} are entropy changes of the universe, surroundings, and the reacting system, respectively. From the second law of thermodynamics,

$$\Delta S_{uni} = \Delta S_{surr} + \Delta S_{sys} \geq 0 \qquad \text{----- (B-2)}$$

Under constant pressure, the change of heat is equal to the change of enthalpy. The heat can be exchanged between the reacting system and surroundings,

$$\Delta H_{sys} = -\Delta H_{surr} \qquad \text{----- (B-3)}$$

Gibbs free energy of the reacting system ΔG_{sys} is defined by

$$\Delta G_{sys} = \Delta H_{sys} - T\Delta S_{sys} \qquad \text{----- (B-4)}$$

Since the change of surrounding is a reversible process,

$$\Delta S_{surr} = \Delta Q_{surr}/T = \Delta H_{surr}/T \qquad \text{----- (B-5)}$$

or $\qquad \Delta H_{surr} = T\Delta S_{surr} \qquad \text{----- (B-6)}$

Substituting Eq. (B-3) and (B-6) into Eq. (B-4) yields

$$\Delta G_{sys} = -T(\Delta S_{surr} + \Delta S_{sys}) = -T\Delta S_{uni} \qquad \text{----- (B-7)}$$

Since T is always positive, $\Delta S_{uni} \geq 0$ is equivalent to $\Delta G_{sys} \leq 0$

Chapter 2
Rates of chemical reactions:
The effect of concentration

Chapter 2   Rates of chemical reactions: The effect of concentration

In this chapter, we discuss the phenomenological description of elementary chemical reactions with the help of rate equations. This approach introduces a method to evaluate the reaction rate constants from the experimental kinetics results. Since the primary purpose is to understand the interfacial reaction mechanism through kinetics, we will not go deeply into the mathematical treatment of kinetic equations.

2.1   Empirical observation

It is generally found that the reaction rates of the chosen species depend on the concentrations of the reactants and products and external parameters, such as temperature. As already mentioned, a standard convention in chemical kinetics is to use the chemical symbol enclosed in brackets for species concentration; thus, [X] denotes the concentration of X. This definition is adopted in this book. [X] can be either a volume concentration (moles/reacting volume) or a surface concentration (moles/reacting surface area) depending on the reacting system.

Consider the following carbon oxidation reaction at a constant temperature of 773K.

$$C + O_2 \rightarrow CO_2 \qquad\qquad \text{----- (2-1)}$$

At this temperature, CO generation is negligibly small. This reaction is carried out in a batch reactor with constant volume, as shown in Fig. 2-1. The reactor is well mixed, and no mass transfer resistance is assumed. Since the amount of carbon is much larger than the amount of O_2, the change of carbon amount can be negligible. In the reaction (2-1), when 1 mol of O_2 decreases, 1 mol of CO_2 increases. Thus, the total pressure does not change with the reaction.

The typical concentration changes of the reactant O_2 and the product CO_2 in the reactor with time are schematically shown in Fig. 2-2.

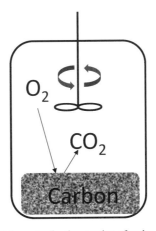

Fig. 2-1   Batch reactor for the reaction of carbon and oxygen.

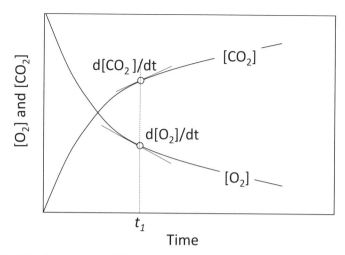

Fig. 2-2   Concentration of CO_2 and O_2 with time in the reaction of $C + O_2 \rightarrow CO_2$.

The rate of changes of concentration of $[O_2]$ and $[CO_2]$ at any time of t_1 are proportional to the slope of its concentration curve, as shown in Fig. 2-2. Thus, the rate of change of the concentration $[O_2]$ can be reasonably defined by $-d[O_2]/dt$. Another convention is to choose negative derivatives for the reactants such as $-d[O_2]/dt$ and positive ones for the products such as $d[CO_2]/dt$ since the reactants decrease and products increase with time in the reaction. Here we use the concentration to define the rate since the concentration is an extensive parameter, and it does not depend on the total amount of the reacting system. Of course, we can use intensive parameters such as the moles of the reacting species. In this case, however, since the rates depend on the total volume or the amount of reacting system and are quite inconvenient, the use of concentrations is preferable.

Since the reaction (2-1) is a heterogeneous reaction, the surface concentration of O_2 should be used instead of the bulk concentration of $[O_2]$. However, the surface concentration is generally proportional to the bulk concentration except in extremely high gas pressure conditions. Thus, it is very convenient to use the bulk concentration instead of the surface concentration because it is not necessary to evaluate the surface area for the reaction. The details of the relation between the bulk and surface concentration are explained based on the Langmuir adsorption isotherm in section 5.3.

When a carbon oxidation reaction with O_2 is carried out at more than 1273 K, the product is CO. The reaction can be expressed by

$$C + O_2 \rightarrow 2CO \qquad\qquad \text{----- (2-2)}$$

In reaction (2-2), the reaction simultaneously consumes 1 mole of O_2 and yields 2 moles of CO. Thus, the production rate of CO becomes twice that of O_2 consumption. Therefore, if we accept the above definition of the rate, such as $-d[O_2]/dt$, each rate of the reactants and products has different values even in the same reaction. This inconvenience is simply solved by introducing the stoichiometric coefficients of the reaction. The new definition of the rates for the reaction (2-1) with stoichiometric coefficients gives,

$$-d[O_2]/dt = (1/2)d[CO]/dt \qquad\qquad ---- (2\text{-}3)$$

If we adopt the new definition of the rate "$(1/2)d[CO]/dt$" instead of "$d[CO]/dt$," the rates of each reactant and product have the same value in the focussed reaction. Although the total pressure gradually increases with reaction time in the reaction (2-2), the apparent rate can be expressed by the slope of its concentration curve.

It has long been customary, however, to refer to $-d[X]/dt$ as the reaction rate of X instead of $-(1/a)d[X]/dt$. "a" is the stoichiometric coefficient of the reaction concerned. As already mentioned, this has the disadvantage of being different for different substances in a reaction. However, still many kinetics studies use the traditional $d[X]/dt$ because of its convenience. In this book, the conventional one is used as the "rate" for most cases since the rates of other reacting species are easily evaluated using the stoichiometric relation.

Experimentally, the reaction rate depends on the concentration of one or more of the reactants and products. Thus, in many cases, the rate is expressed by combining these concentrations at a constant temperature.

$$d[A]/dt = f([A],[B],[P],[Q],\ldots\ldots) \qquad\qquad ----- (2\text{-}4)$$

and this expression is called the *rate law* for the reaction. The rate law equation might be a complicated function of the concentrations (such as the rate of HBr formation in (1-4)). However, the rate law is often expressed as a simple product of the concentrations of species each raised to some power, such as

$$d[A]/dt = k[A]^m[B]^n[P]^p[Q]^q\ldots\ldots \qquad\qquad ----- (2\text{-}5)$$

where m, n, p, q ... are constants, and k is a proportional constant known as the *reaction rate coefficient* (or *rate constant*). The exponents m, n, p, q .... are called the partial orders of reaction concerning A, B, and P, Q ...., and the sum of the partial orders, $m + n + p + q$..., gives the overall order of the reaction.

The proportional constant k in the equation (2-5) is properly termed the rate coefficient rather than the rate constant. The latter term should be reserved for the coefficient in rate expression for elementary reactions.

It is noted that the *rate law and order of a reaction is a purely experimental concept*. There is no relation between the stoichiometric coefficients of a reaction and the reaction order concerning the various reacting substances except elementary reactions. Thus, the order of a reaction is not necessarily an integer: it can be any real number or fraction. The concept of the order of the reaction is applicable when the rate law is expressed as a simple product of the concentrations of species, with each raised to some power.

As already mentioned, the rate of HBr formation by the reaction between H_2 and Br_2 is given by

$$\frac{d[HBr]}{dt} = \frac{2k_2 \left(\frac{k_3}{k_4}\right)\left(\frac{k_1}{k_5}\right)^{\frac{1}{2}} [H_2][Br_2]^{\frac{1}{2}}}{k_3/k_4 + [HBr]/[Br_2]} \qquad \text{----- (1-4).}$$

In this case, we cannot determine the order of a reaction for this reaction. In other words, the order of the reaction ***never*** tells us how the reaction takes place. The precise relationship between the reaction rate and concentration of the reactants is an empirical fact and cannot be guessed from stoichiometry. It therefore must be determined experimentally.

2.2 Elementary and complex reactions

Thus far, we have defined the reaction rate in terms of concentrations, orders of reaction, and reaction rate coefficients. As described in its differential form, the rate law shows how the reaction rate depends on the concentrations. It is often useful to know how the concentrations themselves vary in time. The time behavior of the concentration is easily developed by integrating the rate law for a particular rate expression.

Most elementary reactions are either (1) first-order (for instance, intermolecular rearrangements or decompositions), or (2) second-order (the reaction proceeds by an interaction between two molecules that give products). The majority of chemical reactions can be understood in terms of a combination of these elementary reactions. Thus, the details of the first-order and second-order reactions are considered. Then, the derivation of the rate equations for complex reactions are briefly described.

In this section, the following two assumptions are applied for the analysis. (1) The reaction volume for homogeneous reactions and reaction surface area for heterogeneous reactions are constant during the reaction. (2) The reactions were carried out in a batch reactor. The concentration [A] is either the volume concentrations or surface concentrations. As discussed later in Section 5.3, the surface concentration is proportional to the bulk concentration at a relatively high temperature.

2.3 Rate equations

2.3.1 The first-order irreversible reaction

Since the first-order reaction is the base of all analyses of the reaction kinetics, the characteristics of this reaction are extensively explained here. The differential form of the rate law for a first-order irreversible reaction A → P (products) is

$$d[A]/dt = -k\,[A]. \qquad \text{----- (2-6)}$$

Let [A(0)] be the initial concentration of A and let [A(t)] be the concentration at time t. k is the reaction rate constant and its unit is 1/time. In the experimentally deduced rate laws, the measured proportional factors are called reaction rate coefficients. As a particular case, it is called rate constant in an elementary reaction.

Integration of (2-6) yields

$$-\ln([A(t)] / [A(0)]) = kt, \qquad \text{----- (2-7)}$$

A plot of $-\ln([A(t)] / [A(0)])$ vs. t, as shown in Fig. 2-3, gives a straight line through the origin, and the slope gradient is equal to k. The units of first-order rate constant k are time^{-1}.

The equation (2-7) can be expressed in various forms, and one of them is the integrated form of (2-7),

$$[A(t)] = [A(0)]e^{-kt}. \qquad \text{----- (2-8)}$$

The rate of the product concentration $[P(t)]$ with time is given by,

$$d[P(t)]/dt = k\,[A(t)] \qquad \text{----- (2-9)}$$

and by substituting $([A(0)] - [P(t)])$ for $[A(t)]$,

$$[P(t)] = [A(0)](1 - e^{-kt}) \qquad \text{----- (2-10)}$$

or

$$-\ln([A(0)] - [P(t)]) + \ln[A(0)] = kt \qquad \text{----- (2-10)}'$$

From Eq. (2-8), the consumption of the reactant in a first-order irreversible reaction is presented by the exponential function with time. It means that *the characteristics of properties of first-order irreversible kinetics reflect the properties of the exponential function.* The following features are of practical interest in the first-order irreversible reaction.

(a) Equation (2-8) is extended to

$$-\ln[A(t)] + \ln[A(0)] = kt, \qquad \text{----- (2-11)}$$

Plots of $-\ln[A(t)]$ versus t with different initial concentrations of $[A(0)]_i$ $(i = 1,2,3)$ are shown in Fig. 2-4. They are parallel straight lines with the same slope gradient equal to k in semi-logarithmic coordinates. It means that the rate constant k is independent of the initial concentrations of $[A(0)]_i$ $(i = 1,2,3)$. In other words, the information of the initial concentration $[A(0)]$ is

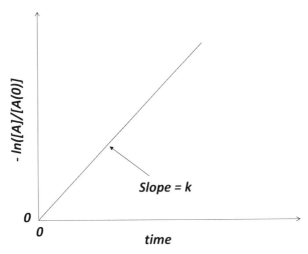

Fig. 2-3   First-order plot.

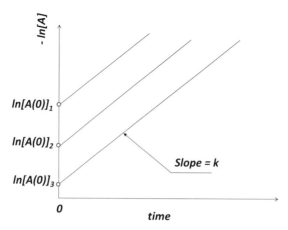

Fig. 2-4   First-order plot with various initial concentrations.

not necessary to determine the rate constant k in a first-order irreversible reaction.

(b) A first-order kinetics curve does not change under the linear transformation of the concentration. This is simply seen by substituting n[A] into [A(t)] in (2-6),

$$d(n[A(t)])dt = -k(n\ [A(t)])\qquad\text{----- (2-12)}$$

n can be cancelled in the equation (2-12) to obtain the same equation (2-6). It means that the kinetic equation remains unaltered by replacing [A] by n[A]. That is, any of the proportional quantities (weight, electrical conductivity, etc.) can be used to evaluate k directly instead of the concentration. The logarithm of such property also changes linearly with time, and the slope is again $-k$.

(c) Plots the [A(t)] / [A(0)] along with natural log of [A(t)]/[A(0)] as a function of time are shown in Fig. 2-5.

The concentration falls to 1/2 of its initial value after a time t ($= \tau_{1/2}$), called the *half-life* of the reactant. It is the time that the concentration falls to half of its value, obtained from,

$$[A(t = \tau^{1/2})]/[A(0)] = 1/2 = \exp(-k\tau^{1/2})$$

$$\tau_{1/2} = \ln(2)/k = 0.693/k.\qquad\text{----- (2-13)}$$

For a first-order reaction, each successive half-life is the same length of time, as shown in Fig. 2-6, and is independent of the starting concentration of [A(t)]. That is, the half-life time of species does not depend on its initial concentration. It is not true for second-order reactions.

This relationship is used in the archaeological field to date the age by measuring ^{14}C content in the findings. ^{14}C changes to ^{14}N due to β decay, and ^{14}C's half-life time is about 5730 years. Since the decay of ^{14}C in the samples is described by the 1st-order reaction, the ^{14}C concentration with time is presented by,

$$[^{14}C(t)] = [^{14}C(0)]\left(\frac{1}{2}\right)^{t/\tau_{1/2}}$$

where $[^{14}C(t)]$ is the ^{14}C concentration in the total carbon contained in the sample at t, $[^{14}C(0)]$

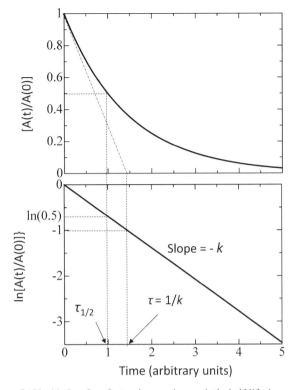

Fig. 2-5 Decay of A[t] with time for a first-order reaction. $\tau_{1/2}$ is the half-life time, and τ is the lifetime of the reaction.

is the initial concentration of ^{14}C, $\tau_{1/2}$ is half-life time, t is a particular time. Since $T(1/2)$ is known to be about 5730 years and $[^{14}C(0)]$ is reasonably estimated by various methods, t can be evaluated from the measurement of $[^{14}C(t)]$. We can use this decay phenomenon as a kind of clock that allows us to peer into the past and determine absolute dates for everything containing carbon, from wood to food and pollen, and so on.

A concept similar to a half-life time is *a lifetime*. The time that the concentration falls to $1/e$ of its initial value is called the *lifetime* of the reactant and is equal to be $1/k$.

$$[A(t = \tau)]/[A(0)] = 1/e = \exp(-k\tau)$$

Then, $\tau = 1/k$ ----- (2-14)

The *lifetime* τ can also be evaluated from the point where the tangent of the concentration gradient at $t = 0$ intersects the horizontal axis, as shown in Fig. 2-5.

As is the case with the half-life time, the *lifetime* τ is determined only by the reaction rate constant and does not depend on the initial concentrations.

What is the meaning of *lifetime* in reactions? If we consider the large number of reactant molecules in a reactor, some will react instantly while others may react after a very long time. What is the average staying time of all these molecules in the reactor before the reaction oc-

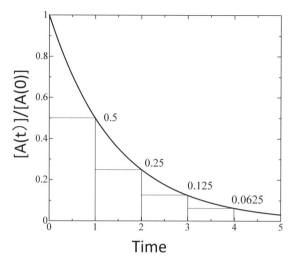

Fig. 2-6 Decay of A[t] for a first-order reaction. Five half-lives for the reaction are shown. For a first-order decay, $\tau_{1/2}$ is independent of the initial concentration on the decay curve.

curs? The number of molecules $N(t)$ that react within t and $t + dt$ in the case of the first-order reaction is calculated by replacing $[A(t)]$ by $N(t)$ as follows;

$$N(t) = N(0)e^{-kt}. \qquad \text{----- (2-15)}$$

Differentiate (2-13),

$$dN(t)/dt = -kN(0)e^{-kt} \qquad \text{----- (2-16)}$$

where $N(0)$ and $N(t)$ is the initial molecular number in the reactor and that after t, respectively. Since a minus sign simply represents the reduction in the number of molecules in the reaction, taking this minus and transforming it yields $dN(t) = kN(0)e^{-kt}dt$.

These $dN(t)$ molecules stayed without reaction for the period of t and then reacted. In other words, the number of molecules that have a waiting time of t before the reaction is $dN(t)$. Thus, the average staying time t_{st} until reaction starts can be calculated by Integrating $tdN(t)$ to infinite time and divides that value by $N(0)$,

$$t_{st} = \frac{\int_0^\infty kte^{-kt}\, dt}{N_0} = \int_0^\infty kte^{-kt}\, dt = 1/k \qquad \text{----- (2-17)}$$

That is, the average staying time t_{st} is equal to $1/k$ or *lifetime* τ. Namely, *lifetime* τ in a reaction means the average waiting time of reactants before they react. The units of the first-order rate constant k are time^{-1}. Thus, when k is 4 sec^{-1}, the average staying time of reacting species before the reaction is 0.25 seconds.

2.3.2   The second-order reaction

There are two types of second-order reactions. Type (a): the reaction between the same atoms or molecules, $A + A \rightarrow P$ (product), and Type (b): the reaction between the different atoms or molecules, A + B → P (product). An example of reaction type (a) is absorbed nitrogen removal from molten iron. The reaction between the adsorbed nitrogen atoms on the molten iron forms the adsorbed nitrogen gas; $2N(ad) \rightarrow N_2(ad)$. $N(ad)$ and $N_2(ad)$ are adsorbed nitrogen atoms and nitrogen gas on the molten iron surface. The example of type (b) is the reaction between the adsorbed oxygen atoms $O(ad)$ and carbon atoms $C(ad)$ on the molten iron to form adsorbed CO gas; $C(ad) + O(ad) \rightarrow CO(ad)$. $C(ad)$ and $O(ad)$ are adsorbed C and O on the molten iron surface. $CO(ad)$ is the adsorbed CO gas on the molten iron surface.

(a) The rate equation for the first type is

$$d[A(t)]dt = -k \, [A(t)]^2 \qquad \text{----- (2.18)}$$

The unit of a second-order rate constant is (concentration)$^{-1}$ time^{-1}. Integration of (2-18) gives,

$$1/[A(t)] - 1/[A(0)] = kt \qquad \text{----- (2-19)}$$

As shown in Fig. 2-7, a plot of $1/[A(t)]$ versus t with different $[A(0)]$s yields straight lines whose intercept is $1/[A(0)]$ and whose slopes are the rate constant k. The different initial concentrations $[A(0)]_i$ (i = 1, 2, 3 ....) result in parallel straight lines, as in the case of a first-order reaction.

Unlike a first-order reaction, second-order kinetics curves change under the linear transformation of the concentration. Thus, the measurements of the exact values of the concentration are required to evaluate the rate constant in the second-order reaction. For example, in a first-order reaction, the rate constant can be directly calculated from the slope of the semi-logarithmic plot of the weight changes with time, and it is not necessary to measure the concentration changes. For second-order reactions, however, the obtained slope from the direct plots of weight changes using eq. (2-19) does not reflect the true rate constant. To evaluate the true rate constant, the concentration changes converted from the weight change must be plotted

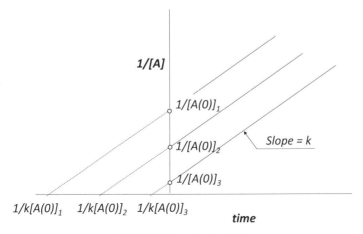

Fig. 2-7 Second-order plot.

instead of the direct plot of weight changes.

The half-life of A is obtained after substituting $[A(0)]/2$ for $[A(t)]$ in eq. (2.19),

$$t_{1/2} = 1/(k\,[A(0)]) \qquad\qquad ----- (2\text{-}20)$$

In contrast to first-order kinetics, the half-life of a second-order reaction varies with the initial concentration.

(b) The rate equation for the second type of reactions that are second-order overall but first order in each of two reactants is

$$d[A(t)]dt = -k\,[A(t)][B(t)] \qquad\qquad ----- (2\text{-}21)$$

It is convenient to treat this reaction type for the following two conditions:

(b-1) Because the initial concentration of B is very much larger than that of A, the concentration of B remains practically unchanged during the reaction. Thus, $[B(t)] \approx [B(0)]$. Then the rate equation is simplified to

$$d[A(t)]dt \approx -k\,[A(t)][B(0)] = -k_{eff}\,[A(t)] \qquad ----- (2\text{-}22)$$

where k_{eff} is an effective rate coefficient and is.

$$k_{eff} = k\,[B(0)]. \qquad\qquad ----- (2\text{-}23)$$

The rate equation is essentially the same as the first-order reaction (2-6). Thus, a reaction that normally has to be analyzed as a 2nd-order reaction can be analyzed as a first-order reaction focusing on [A] when $[B(0)] \gg [A(0)]$. The values of k_{eff} do not depend on $[A(0)]$ since a first-order rate coefficient k_{eff} is concentration-independent.

(b-2) For the case that $[A(0)]$ are comparable to $[B(0)]$: We must exactly solve the differential equation of (2-21). The integration of (2.21) gives,

$$\frac{1}{([A(0)] - [B(0)])}\, ln\, \frac{[B(0)][A(t)]}{[A(0)][B(t)]} = kt \qquad ------ (2\text{-}24)$$

Details of the deduction of (2-22) can be found elsewhere [1]. A graph of the left-hand term versus t yields a straight line of slope equal to k.

Obtaining the reaction rate coefficient by applying eq. (2-24) is experimentally very demanding since many experiments must be carried out with varying $[A(0)]$ and $[B(0)]$. The experimental technique to meet the reactions of type (b-2), set the initial concentrations of A and B equal ($[A(0)] = [B(0)]$), then, the rate equation (2-19) becomes the same as the equation (2-16), and is easily integrated.

2.4 Complex reactions and their reaction mechanisms

Many reactions consist of combinations of first-order or second-order reactions, and these are called complex reactions. In this section, we will take typical complex reactions and explain their characteristics.

2.4.1 The reversible first-order reaction

A reversible reaction consists of two reaction steps in the opposite direction. Here, we consider a reversible first-order ***elementary*** reaction A $\rightleftarrows$ P, with the rate constant k_F in the forward direction (A → P) and k_B reverse direction (P → A). The rate equation is given by,

$$-d[A(t)]/dt = d[P(t)]/dt = k_F[A(t)] - k_B[P(t)] \qquad ----- (2\text{-}25)$$

As already mentioned in chapter 1, the forward and opposing reaction rates are equal at equilibrium,

$$k_F[A]_{eq} - k_B[P]_{eq} = 0 \qquad ----- (2\text{-}26)$$

where $[A]_{eq}$ and $[P]_{eq}$ are the equilibrium concentration of A and P, respectively. k_F and k_B can be correlated through the equilibrium constant K_{eq} of the reaction A $\rightleftarrows$ P.

$$K_{eq} = [P]_{eq} / [A]_{eq} = k_F / k_B \qquad ----- (2\text{-}27)$$

It is noted that the equation (2-27) can be applied only under the equilibrium condition. However, if both forward and backward reactions are elementary reactions, (2-27) holds even when the equilibrium condition is not met.

Using the mass balance of $[P]_{eq} = ([A(0)] - [A]_{eq})$ and (2-27),

$$[A]_{eq} = \{k_B /(k_F + k_B)\} [A(0)] \qquad ----- (2\text{-}28)$$
$$[P]_{eq} = \{k_F /(k_F + k_B)\} [A(0)] \qquad ----- (2\text{-}29)$$

Replacing [P(t)] in (2-25) by ([A(0)] – [A(t)]) and rearrange to get

$$d[A(t)]/dt = - (k_F + k_B)\{ [A(t)] - k_B[A(0)]/(k_F + k_B)\} \qquad ----- (2\text{-}30)$$

The second term in the curly bracket is simply $[A]_{eq}$ according to (2-28). Hence

$$d[A(t)]/dt = - (k_F + k_B)([A(t)] - [A]_{eq}) \qquad ----- (2\text{-}31)$$

This is easily integrated to

$$ln \frac{[A(t)] - [A]_{eq}}{[A]_0 - [A]_{eq}} = -(k_F + k_B)t \qquad ----- (2\text{-}32)$$

Therefore, a graph of $ln([A(t)] - [A]_{eq})$ versus t gives a straight line of slope $-(k_F + k_B)$. Individual k_F and k_B can be obtained from the measured value of the slope $(k_F + k_B)$ combined with the equilibrium constant of K_{eq} $(= k_F/k_B)$.

When the forward rate constant is much larger than the reverse one, $k_F \gg k_B$, the equilibrium constant K_{eq} has a large value, $K_{eq} \gg 1$. Under the equilibrium condition with $K_{eq} \gg 1$, A will react almost to all B. This situation is schematically shown in Fig. 2-8 (a).

In this case, $k_F + k_B \approx k_F$, and $[A]_{eq} \approx 0$, so that (2-32) reduces to

$$-\ln([A(t)]/[A(0)]) = k_F t, \qquad ----- (2\text{-}33)$$

This rate equation is simply the same as (2-7). It should be noted that it is very hard to tell the difference between the first-order reversible reaction with a large equilibrium constant ($K_{eq} \gg 1$) and a simple first-order irreversible reaction from the measurements of the concentration changes with time.

In the case of $K_{eq} \ll 1$, or $k_F \ll k_B$, the term of $k_B/(k_F + k_B)$ and $k_F/(k_F + k_B)$ in (2-24) can be approximated to be unity and zero, respectively. Then,

$$[A]_{eq} = \{k_B /(k_F + k_B)\}[A(0)] \approx [A(0)] \qquad ----- (2\text{-}34)$$
$$[P]_{eq} = \{k_F /(k_F + k_B)\} [A(0)] \approx 0 \qquad ----- (2\text{-}35)$$

34

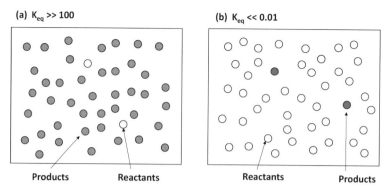

(a) K$_{eq}$ >> 100

(b) K$_{eq}$ << 0.01

Products   Reactants   Reactants   Products

Fig. 2-8 Expected composition of an equilibrium mixture under the conditions of (a) K$_{eq}$ >> 100, and (b) K$_{eq}$ << 0.01.

In this case, the backward reaction rate is much faster than the forward one: the generated P immediately returns to A again. Thus, the reaction does not seem to progress under the condition of K$_{eq}$ << 1. The equilibrium situation with K$_{eq}$ << 1 is schematically illustrated in Fig. 2-8 (b). Even under this condition, however, the forward and backward reactions always occur dynamically even when the reactions are apparently at a standstill.

Meanings of the positive and negative values of ΔG^0

In section 1.3, it is already shown that the sign of ΔG (the Gibbs free energy of the system) but not of ΔG^0 (Gibbs standard free energy change) indicates whether the reaction will occur or not. Then, what is the meaning of the sign of ΔG^0?

K$_{eq}$ is expressed by using ΔG^0 (standard Gibbs free energy change),

$$\Delta G^0 = -RT \, ln \, K_{eq} \qquad ----- (2\text{-}36)$$

Thus, the reaction with negative ΔG^0 (ΔG^0 < 0) means K$_{eq}$ > 1, and that with a positive one means K$_{eq}$ < 1. In other words, the reaction with a large positive ΔG^0 corresponds to the equilibrium condition of (b), and a large negative ΔG^0 corresponds to the equilibrium condition of (a) shown in Fig. 2-8. When K$_{eq}$ is more than 10^4, the equilibrium state consists of almost 100% products, while K$_{eq}$ is less than 10^{-4}, consisting of almost 100% reactants. Using the eq. (2-36), the value of ΔG^0 will be −23 kJ/mol when K$_{eq}$ = 10^4, and 23 kJ/mol with K$_{eq}$ = 10^{-4}. In other words, when |ΔG^0| becomes more than 23 kJ/mol, which is not such a large value, the equilibrium state consists of almost 100% reactants or 100% products.

The equilibrium state with a large negative value of ΔG^0 consists of a significant quantity of products and a small amount of reactant. It is noted that ΔG^0 is relevant only for the equilibrium state, and it is entirely irrelevant for the time required to reach that equilibrium state. Reactions with a large negative value of ΔG^0 do not necessarily take just a short time to reach equilibrium. They may reach equilibrium almost momentarily or take a million years. In our experience, it is generally found that reactions with a large negative ΔG^0 are likely to occur.

But there is no theoretical basis for this tendency. Therefore, experiments must confirm whether a reaction occurs at a reasonable rate or not.

2.4.2 Series of two first-order reactions

Another important complex reaction mechanism is that of consecutive reactions. In these types of reactions, reactant A is converted to product P via an intermediate product B.

$$A \rightarrow B \qquad\qquad ----- (2\text{-}37)$$
$$B \rightarrow P \qquad\qquad ----- (2\text{-}38)$$

For heterogeneous interfacial reactions, it is assumed that each reaction (2-37) and (2-38) is an elementary reaction and is carried out with **the same constant reaction surface area.** For these reactions, the rate equations are

$$d[A(t)]/dt = -k_1[A(t)] \qquad\qquad ----- (2\text{-}39)$$
$$d[B(t)]/dt = k_1[A(t)] - k_2[B(t)] \qquad\qquad ----- (2\text{-}40)$$
$$d[P(t)]/dt = k_2[B(t)] \qquad\qquad ----- (2\text{-}41)$$

where k_1 and k_2 are the rate constant for the reaction (2-37 and (2-38) respectively.

Solving these differential equations with the initial concentrations of A is [A(0)] and [B(0)] = [C(0)] = 0,

$$[A(t)] = [A(0)] \exp(-k_1 t). \qquad\qquad ----- (2\text{-}42)$$

$$[B(t)] = \frac{k_1}{k_2 - k_1} [A(0)](e^{-k_1 t} - e^{-k_2 t}) \qquad\qquad ----- (2\text{-}43)$$

$$[P(t)] = [A(0)] - [B(t)] - [A(t)]$$

$$= [A(0)]\left\{1 + \frac{1}{k_1 - k_2}(k_2 e^{-k_1 t} - k_1 e^{-k_2 t})\right\} \qquad\qquad ----- (2\text{-}44)$$

The details of the deduction of these equations can be found elsewhere [2].

Approximate solutions

The equations (2-43) and (2-44) are complex, but they can be simplified based on specific reaction conditions.

Case 1: $k_1 \gg k_2$ (The reaction A $\rightarrow$ B is faster than B $\rightarrow$ P). The equation (2-43) is reduced to

$$[B(t)] = [A(0)](e^{-k_2 t} - e^{-k_1 t}) \qquad\qquad ----- (2\text{-}45)$$

The second term in the parentheses quickly approaches zero, while the first term remains near unity. Consequently, the concentration of B rapidly reaches a value close to [A(0)] and then [B] is approximately expressed by

$$[B(t)] \approx [A(0)](e^{-k_2 t}) \qquad\qquad ----- (2\text{-}46)$$

and approximated [P(t)] is

$$[P(t)] \approx [A(0)](1 - e^{-k_2 t}) \qquad\qquad ----- (2\text{-}47)$$

Note that [B(t)] and [P(t)] changes with time in the same manner as [A(t)] and [P(t)] for the

36

first-order reaction shown in (2-8) and (2-10). That is, A is very rapidly consumed at the early stage and P production proceeds as if it starts from B after the short induction time. The lifetime of A (τ_A) and B (τ_B) are $1/k_1$ and $1/k_2$, respectively. The lifetime τ in a reaction means the average staying time of reactants before they react. Since $k_1 \gg k_2$, $\tau_A \ll \tau_B$. In other words, A's lifetime is very short compared with that of B, or A is consumed more rapidly than B.

The variation in concentration of A, B, and P with time is represented in Fig. 2-9 for the case of $k_1 = 10k_2$. As shown in Fig. 2-9, the substantial conversion of A to B occurs before appreciable P is produced.

Under the condition of $k_1 \gg k_2$, the concentration equation of (2-47) includes only k_2 but not k_1. It means that the reaction of B → P practically determines the concentration changes of [P(t)] with time, and the reaction step of A → B does not affect the production of P. Surprisingly, even the reaction A → B is controlled by the reaction B → P except at the early stage of the reaction since the concentration of [B(t)] with time does not contain k_1. Namely, the reaction of B → P dominantly controls the overall reaction and is called the **reaction controlling step or rate control step** in the series reaction.

Case2: $k_2 \gg k_1$ (The reaction B → P is faster than A → B). The solution [B(t)] given by eq. (2-44) reduces to

$$[B(t)] \approx \frac{k_1}{k_2}[A(0)](e^{-k_1 t} - e^{-k_2 t}) \qquad ----- (2\text{-}48)$$

The second term in the parentheses rapidly approaches zero since $k_2 \gg k_1$, while the first term is still close to unity. Consequently, [B(t)] rapidly approaches $(k_1/k_2)[A(0)]$ at the early stage and then start to decay according to $(k_1/k_2)[A(0)] \exp(-k_1 t)$. Because k_1/k_2 is very small, or $k_1/k_2 \approx 0$, [B(t)] is much less than [A(0)] even at the maximum concentration of [B(t)] and

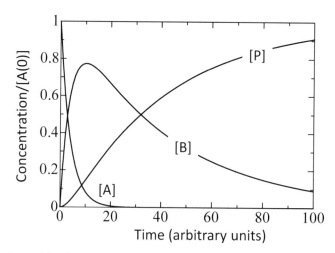

Fig. 2-9 Composition-time curves for a series of two first-order reactions with $k_1 \gg k_2$, ($k_1 = 0.25$, $k_2 = 0.025$).

can be assumed to be almost zero.

Then, under the condition of $k_2 \gg k_1$, [A(t)], [B(t)], [P(t)] are reasonably expressed by

$$[A(t)] = [A(0)] \exp(-k_1 t) \qquad \text{----- (2-49)}$$
$$[B(t)] \approx (k_1/k_2) [A(0)] \exp(-k_1 t) \approx 0 \qquad \text{----- (2-50)}$$
$$[P(t)] \approx [A(0)] (1 - \exp(-k_1 t)) \qquad \text{----- (2-51)}$$

The behaviors of both [A] and [P] are given by simple first-order reaction curves, except for the time-lag of the formation of P. The rate equation (2-51) does not contain k_2. It means that the reaction step of A → B determines the concentration changes of [P(t)] even when P is produced from B. In this case, the reaction controlling step in the series reaction is A → B. The reaction of A → B → P proceeds like the reaction of A → P without the formation of intermediate B. From the equation of (2-49) and (2-50), [A] + [P] ≈ [A(0)]. It means that the increase of [P] almost matches the decrease of [A]. This behavior is presented in Fig. 2-10.

For the reaction with the condition of $k_2 \gg k_1$, if we pay attention only to the early stage and final stage of the reaction, we may overlook the existence of the reactive intermediate B in the reaction since the amount of the intermediate B is too small to measure. Intermediates such as free radicals, ions, and energy-excess intermediates have been recently found due to the developments of analytical instruments.

2.4.3   Parallel reactions

A typical simple parallel reaction as competing two first-order reactions is

$$A \rightarrow B \qquad \text{----- (2-52)}$$
$$A \rightarrow C \qquad \text{----- (2-53)}$$

The rate laws for these reactions are

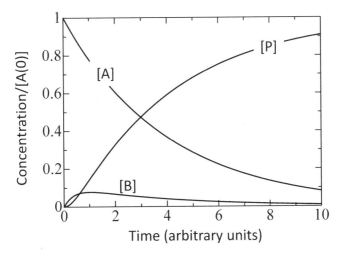

Fig. 2-10   Composition-time curves for a series of two first-order reactions $k_2 \gg k_1$, ($k_1 = 0.25$, $k_2 = 2.5$).

$$d[A(t)]/dt = -k_1[A(t)] - k_2 [A(t)] = -(k_1 + k_2) [A(t)] \qquad \text{----- (2-54)}$$

$$d[B(t)]/dt = k_1[A(t)] \qquad \text{----- (2-55)}$$

$$d[C(t)]/dt = k_2[A(t)] \qquad \text{----- (2-56)}$$

where k_1 and k_2 are the reaction rate constants of (2-46) and (2-47), respectively.

Solving these equations gives,

$$[A(t)] = [A(0)] \exp \{-(k_1 + k_2)t\} \qquad \text{----- (2-57)}$$

$$[B(t)] = [k_1/(k_1 + k_2)][A(0)](1 - \exp \{-(k_1 + k_2)t\}) \qquad \text{----- (2-58)}$$

$$[C(t)] = [k_2/(k_1 + k_2)][A(0)](1 - \exp \{-(k_1 + k_2)t\}) \qquad \text{----- (2-59)}$$

From eqs. (2-58) and (2-59), the ratio of $[B(t)]/[C(t)]$ is k_1/k_2 and is constant at all times. The ratio k_1/k_2 is called the **branching ratio**. Using the slope $(k_1 + k_2)$ obtained from eq. (7-59) and k_1/k_2 measured from $[B(t)]/[C(t)]$ at a particular time, k_1, and k_2 are calculated.

When $k_1 \gg k_2$, $[B(t)]$ and $[C(t)]$ are approximated to

$$[B(t)] \approx [A(0)](1 - \exp(-k_1 t)) \qquad \text{----- (2-58)}'$$

$$[C(t)] \approx 0 \qquad \text{----- (2-59)}'$$

And when $k_1 \ll k_2$, $[B(t)]$ and $[C(t)]$ are approximated to

$$[B(t)] \approx 0 \qquad \text{----- (2-58)}''$$

$$[C(t)] = [A(0)](1 - \exp(-k_2 t)) \qquad \text{----- (2-59)}''$$

That is, in a parallel reaction, product formation is controlled by a faster reaction. In an industrial process, undesirable reactions must be suppressed. For this purpose, a catalyst is used. **The role of the catalyst is not only the enhancement of the reaction rate but also the selectivity of the reaction**. In other words, the appropriate catalyst enhances the specified reaction rate to change the branching ratio. It is noted that the rate to reach the equilibrium state varies depending on the catalysts, but the composition in the equilibrium state does not change according to the catalyst. The reaction profile for the reaction A → B with and without a catalyst is shown in Fig. 2-11. G_1, G_2 are the Gibbs free energies of reactant A and product B, respectively. Q is the activation energy for reaction A → B without a catalyst, and Q_C is that with a catalyst. As shown in Fig. 2-11, the activation energy with a catalyst is much smaller than that without a catalyst. As a result, the reaction rate with a catalyst becomes more significant than that without a catalyst. As shown in Fig. 2-11, the catalyst does not affect the change of free energy.

Thermodynamics of the parallel reaction

Unfortunately, there are no appropriate examples of parallel reactions in the Fe-C-O system. Here, as an example, the organic reaction is shown instead. Many parallel reactions take place in the field of industrial organic synthesis processes. Let's think about the ethylene oxidation reaction. The following three parallel reaction paths can be considered.

I. $C_2H_4 + 3O_2 \rightarrow 2CO_2 + 2H_2O$ \qquad ----- (2-60)

$\Delta G = -1315.0$ kJ/mol \qquad ----- (2-60)'

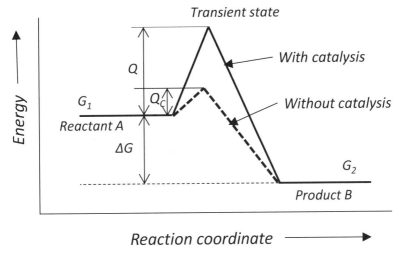

Fig. 2-11   The reaction profile for reaction A → B with and without catalyst. G_1, G_2 are the Gibbs free energies of reactant A and product B, respectively. Q and Q_c are the activation energy for reaction A → B without and with a catalyst.

ΔG (kJ/mole) is the value obtained by subtracting the sum of the free energies of the raw material system consisting of ethylene and oxygen from the sum of the free energies of the production system consisting of CO_2 and H_2O at 298.15 K. ΔG is calculated based on 1 mole of ethylene. The same applies to ΔG in the following reaction.

II. $C_2H_4 + O_2/2 \rightarrow C_2H_4O$ ----- (2-61)

$\Delta G = -81.3$ kJ/mol   ----- (2-61)'

III. $C_2H_4 + O_2/2 \rightarrow CH_3CHO$ ----- (2-62)

$\Delta G = -200.9$ kJ/mol   ----- (2-62)'

C_2H_4O and CH_3CHO are ethylene oxide and acetaldehyde, respectively. More oxidations of C_2H_4O and CH_3CHO produce CO_2 and H_2O. The final products of ethylene oxidation are CO_2 and H_2O.

IV. $C_2H_4O + (5/2)O_2 \rightarrow 2CO_2 + 2H_2O$ ----- (2-63)

$\Delta G = -1233.7$ kJ/mol   ----- (2-63)'

V. $CH_3CHO + (5/2)O_2 \rightarrow 2CO_2 + 2H_2O$ ----- (2-64)

$\Delta G = -1114.1$ kJ/mol   ----- (2-64)'

Thermodynamically, the most stable equilibrium state due to the oxidation of ethylene is the production of CO_2 and H_2O by the direct combustion of ethylene. This is because the decrease in ΔG of this change (the reaction I) is the largest, and the free energy of the generated system is the smallest.

The Gibbs free energy changes of these reactions are schematically shown in Fig.2-12. The diagram indicates that the free energy decreases not only in the reaction that produces CO_2 and H_2O (reaction I) but also in the reaction that produces ethylene oxide (reaction II) and

40

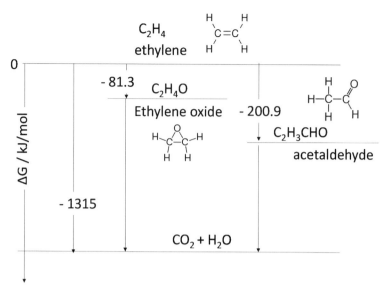

Fig. 2-12 The schematic diagram of Gibbs free energy changes of the ethylene oxidation at 298.15 K.

acetaldehyde (reaction III). Since the Gibbs free energy decreases in any of these reactions, the reactions of II and III can also proceed.

If the largest free energy change system (the reaction I) governs the progress of the ethylene oxidation reaction, we could only make CO_2 and H_2O from ethylene and oxygen. However, the industrial ethylene oxidation process routinely produces ethylene oxide and acetaldehyde from the partial oxidation of ethylene. This result indicates that the reaction does not proceed in the direction in which the Gibbs free energy decreases most but goes in many directions according to where the free energy decreases. That is, the magnitude of ΔG has nothing to do with the reaction rates nor does it indicate **which reaction preferentially proceeds**. The catalyst can be used to efficiently produce the desired product.

2.4.4 Series of two first-order reactions: First step is reversible

Another complex reaction is the combination of the reversible reaction of A $\rightleftarrows$ B and a first-order irreversible reaction of B → P such as A $\rightleftarrows$ B → P. This kind of reaction system occurs quite often. The set of differential equations is given by

$$d[A(t)]/dt = -k_F [A(t)] + k_B[B(t)] \qquad ----- (2\text{-}65)$$
$$d[B(t)]/dt = k_F [A(t)] - (k_B + k_2)[B(t)] \qquad ----- (2\text{-}66)$$
$$d[P(t)]/dt = k_2[B(t)] \qquad ----- (2\text{-}67)$$

where k_F and k_B are the rate constants of the forward and backward reaction of the reversible reaction A $\rightleftarrows$ B, respectively. k_2 is the rate constant of the reaction B → P. The establishment of the exact solutions of these simultaneous equations is quite demanding. Thus, our focus is on the following limiting cases.

Case 1: $k_2 \gg (k_F, k_B)$

In this case, the intermediate B produced from A is quickly converted to P before it reverses and goes back to A. Thus, the concentration of [B(t)] is practically negligibly small during the reaction. Therefore, the concentration changes of [A(t)], [B(t)], and [P(t)] with time are approximately the same with the case of consecutive reactions with $k_2 \gg k_1$. (see Fig. 2-10) This is easily understood since the backward reaction B → A negligibly occurs under $k_2 \gg k_B$.

Case 2: $(k_F, k_B) \gg k_2$

When forward and backward reactions are fast in the reversible reaction of A ⇌ B, and the rate of B → P is slow.

$$A \rightleftarrows B \quad \text{fast}$$

$$B \rightarrow P \quad \text{slow}$$

Under this condition, we may reasonably assume the reversible reaction of A ⇌ B will be at equilibria. Then,

$$[P(t)] = [A(0)] \left(1 - exp\left[-k_{eff} t\right]\right) \quad \text{----- (2-68)}$$

$$[B(t)] = \frac{K_{eq}}{1 + K_{eq}} [A(0)] \, exp\left[-k_{eff} t\right] \quad \text{----- (2-69)}$$

$$[A(t)] = \frac{1}{1 + K_{eq}} [A(0)] \, exp\left[-k_{eff} t\right] \quad \text{----- (2-70)}$$

where $k_{eff} = k_2 K_{eq}/(1 + K_{eq})$.

Under the conditions of $(k_F, k_B) \gg k_2$ and $K_{eq} \gg 1$, the rate equations of (2-62), (2-63) and (2-64) can be expressed by

$$[P(t)] \approx [A(0)] \left(1 - exp\left[-k_2 t\right]\right) \quad \text{----- (2-68)}'$$
$$[B(t)] \approx [A(0)] \, exp\left[-k_2 t\right] \quad \text{----- (2-69)}'$$
$$[A(t)] \approx [A(0)] \, exp\left[-k_2 t\right] \quad \text{----- (2-70)}'$$

These equations do not contain K_{eq}. Thus, the equilibrium constant of K_{eq} has a negligible effect on each rate.

While under the conditions of $(k_F, k_B) \gg k_2$ and $K_{eq} \ll 1$, each concentration change is expressed by

$$[P(t)] \approx [A(0)] \left(1 - exp\left[-k_2 K_{eq} t\right]\right) \quad \text{----- (2-68)}''$$
$$[B(t)] \approx K_{eq} [A(0)] \, exp\left[-k_2 K_{eq} t\right] \quad \text{----- (2-69)}''$$
$$[A(t)] \approx [A(0)] \, exp\left[-k_2 K_{eq} t\right] \quad \text{----- (2-70)}''$$

All rate equations contain K_{eq}. That is, the rates of [A], [B], and [P] are affected by the equilibrium constant K_{eq}. In other words, the reversible reaction A ⇌ B influences the rate of change of each component and is called a ***rate-determining equilibrium reaction***.

Since the rate of forward and backward reactions are fast, [A(t)] and [B(t)] are close to the equilibrium concentrations at the early stage of the reaction. Under the near-equilibrium con-

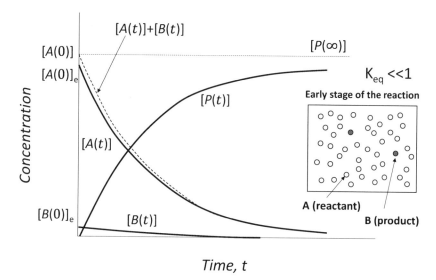

Fig. 2-13 Concentration-time curves for the reaction A $\rightleftarrows$ B $\rightarrow$ P, under the condition $K_{eq} \ll 1$. $[A(0)]_e$ and $[B(0)]_e$ are the equilibrium values of [A] and [B]. The early stage of the reaction is also shown in Fig. 2-13. At the early stage, the amount of [P(t)] is negligibly small.

dition, $[B(t)] \approx [A(t)]/K_{eq}$, and [B(t)] is considerably smaller than [A(t)] during the reaction since $K_{eq} \ll 1$.

Thus, **the reaction seems to proceed from A to directly P**. The concentration changes with time for the condition of $(k_F, k_B) \gg k_2$ and $K_{eq} \ll 1$ are schematically illustrated in Fig. 2.13.

Further reading

R. Schmid R and V. N. Sapunov: Non-formal kinetics, Verlag Chemie GmbH, Weinheim, 1982.

R. S. Berry, S. A. Rice and J. Ross: Physical and Chemical Kinetics, Oxford Uni. Press, Oxford, 2002.

O. Levenspiel: Chemical Reaction Engineering 3rd ed., John Wily & Sons, Hoboken, 1999.

M. J. Pilling and P. W. Seakins, Reaction Kinetics, Oxford Uni. Press, Oxford, 2005.

P. L. Houston: Chemical Kinetics and Reaction Dynamics, McGraw-Hill, 2000.

J. I. Steinfeld, J. S. Francisco, and W. L. Hase: Chemical Kinetics and Dynamics 2nd ed., Prentice-Hall, New Jersy, 1998.

References

[1] P. L. Houston: Chemical Kinetics and Reaction Dynamics, McGraw-Hill, 2000, 43-44.

[2] P. L. Houston: Chemical Kinetics and Reaction Dynamics, McGraw-Hill, 2000, 56-59

Chapter 3
Dependence of reaction rate on temperature and pressure

Chapter 3   Dependence of reaction rate on temperature and pressure

3.1   Effect of temperature on reaction rates

The previous section, it was shown that the concentration dependence on the rate provides a possible reaction mechanism. In the same way, we may get important information about the reaction mechanism from the temperature dependence of the rate. For example, for the reaction $CO(g) + O_2/2(g) \rightarrow CO_2(g)$ catalyzed by nickel oxide, the rate laws show a complex dependence of reacting gas concentrations depending on temperature, as shown in Table 3.1.

The different adsorption behaviors at different temperatures simply cause these different reaction mechanisms. In these ways, a kind of information about the reaction mechanism may be obtained from the temperature dependencies. Details of the adsorption process are discussed in Chapter 5.

The rate expressions are often simple functions of reactant concentrations with a characteristic rate coefficient k. This rate coefficient should be independent of concentrations and time. It does, however, depend strongly on temperature. It is found experimentally that a great majority of reactions has a rate coefficient k that follow the relationship;

$$k = A_p \exp(-E_a/RT) \qquad ----- (3\text{-}1)$$

A_p is called the pre-exponential or frequency factor, and E_a is the activation energy. R is the ideal gas constant (approximately 8.314 J mol^{-1}K^{-1}); T is the absolute temperature. A_p and E_a are constants and can only be determined and defined experimentally.

The relationship (3-1) is known as the Arrhenius equation [1], and a plot of ln k vs. $1/T$ is called an Arrhenius plot. E_a can be deduced by plotting *ln k* as a function of $1/T$ (Arrhenius plot) if a straight line is endorsed in the concerned temperature range. The slope of the Arrhenius plot is $-E_a/R$. Since pre-exponential A_p is independent of temperature, the substitution of the Arrhenius equation for the rate coefficient of k_1 at T_1 from that of k_2 at T_2 yields,

$$ln(k_2/k_1) = -(E_a/R)[1/T_2 - 1/T_1] \qquad ----- (3\text{-}2)$$

When the rate coefficients at different temperatures are known, E_a is easily evaluated by applying eq. (3-2). For the elementary reactions, the measured activation energy has a constant value. However, the activation energies for complex reactions are not necessarily constant. Sometimes, the activation energy changes when the rate-controlling step changes with

Table 3.1   The rate laws of $CO(g) + O_2/2(g) \rightarrow CO_2(g)$ on NiO with temperature, and k_i (i = 1 to 4) is an apparent rate coefficient of the reaction of $CO(g) + O_2/2(g) \rightarrow CO_2(g)$ at each temperature.

Temperature (°C)	Rate law
25-60	$k_1[O_2]$
100-180	$k_2[CO]^{1/2}[O_2]^{1/2}$
200-250	$k_3[CO][O_2]^{1/2}$
250-450	$k_4[CO]$

temperature, as already mentioned.

To understand the temperature dependence of rate coefficient, several theoretical approaches have been considered, including the collision theory, the theory of absolute reaction, and the transition state theory. All of these theories lead to the following temperature dependency of the rate coefficient k.

$$k = A_p T^m \exp(-E_a/RT) \qquad ----- (3\text{-}3)$$

The Eq. (3-3) is similar to (3-1) except for T^m. Generally, the value of m in Eq. (3-3) is in the range of $0 < m < 1$. Thus, the influence of T^m term on the rate coefficient k is negligibly small compared with that of $\exp(-E_a/RT)$. Consequently, the equation (3-1) is a good practical approximation of (3-3) for most reactions.

In general, the temperature variation of the activation energy and pre-exponential term can be small, and an elementary reaction essentially behaves like a straight line in the Arrhenius coordinates. The scattering of the experimental points may well mask slight deviations from linearity. Dominant nonlinear behavior indicates that the rate constant employed is complex or composite related to more than one reaction step.

3.2   Reactions do not follow the Arrhenius equation

In the majority of reactions, the reaction rate coefficients exhibit Arrhenius behavior. That is, the reaction rates increase with temperature. However, in reality, the rate increases with temperature in some cases, but not necessarily according to Arrhenius's law, and in some cases, it even decreases.

Arrhenius plot gives a concave curve downward

The apparent rate at high temperatures is faster than expected from those in the lower temperature range, as shown in Fig. 3-1.

This type of temperature dependency of rate coefficient k typically occurs for parallel reactions. A parallel reaction is that a species can react by more than one pathway. A typical simple parallel reaction as competing for two first-order reactions is

$$A \rightarrow B \qquad ----- (3\text{-}4)$$
$$A \rightarrow C \qquad ----- (3\text{-}5)$$

As already explained (section 2.4.3), the decrease rate of [A] in the above parallel reaction is expressed by

$$d[A(t)]/dt = -k[A(t)] = -k_1[A(t)] - k_2[A(t)] = -(k_1 + k_2)[A(t)] ----- (3\text{-}6)$$

where k_1 and k_2 are the reaction rate constants of (3-4) and (3-5) and k is the overall apparent rate coefficient of [A] decreasing reaction, and $k = k_1 + k_2$. In this case, the Arrhenius equation can be expressed by

$$k = k_1 + k_2 = A_1 \exp[-(E_a)_1/RT] + A_2 \exp[-(E_a)_2/RT] \qquad ----- (3\text{-}7)$$

where A_1 and A_2 are a pre-term factor of the Arrhenius equation, and $(E_a)_1$ and $(E_a)_1$ are activa-

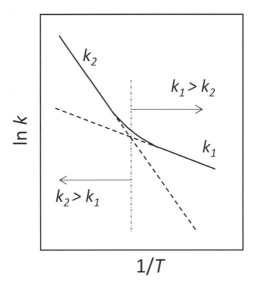

Fig. 3-1   Arrhenius plot with a concave curve downward

tion energies for reaction (3-4) and (3-5). In the above parallel reaction, assume that the reaction A → B preferentially occurs ($k_1 \gg k_2$) in the low-temperature range, while A →C preferentially occurs ($k_1 \ll k_2$) in the high-temperature range. In these conditions, the faster reaction is dominantly observed in all ranges of temperature.

At the low temperature range ($k_1 \gg k_2$): $k \approx k_1 = A_1 \exp[-(E_a)_1/RT]$

At the high temperature range ($k_1 \ll k_2$): $k \approx k_2 = A_2 \exp[-(E_a)_2/RT]$

Therefore, the apparent activation energy will be $(E_a)_1$ in the high-temperature range and $(E_a)_2$ in the low-temperature range.

Arrhenius plot gives a concave curve upward

The apparent rate at high temperatures is slower than expected from those in the lower temperature range, as shown in Fig. 3-2. This type of temperature dependency of k is often found for series reactions.

Suppose the following series 1st order reaction.

A → B,   ----- (3-8)

B → P   ----- (3-9)

Assume k_1 and k_2 are the rate coefficient of (3-8) and (3-9), respectively. A slower reaction controls the production rate of P. In the range of $k_1 \ll k_2$, the rate is controlled by reaction A → B. In the range of $k_1 \gg k_2$, reaction B→ P controls the overall reaction rate. Therefore, if $k_1 \gg k_2$ in the low-temperature range and $k_1 \ll k_2$ in the high-temperature range, the apparent reaction rate can be approximately expressed by

At the low temperature range ($k_1 \gg k_2$): $k \approx k_2 = A_2 \exp[-(E_a)_2/RT]$

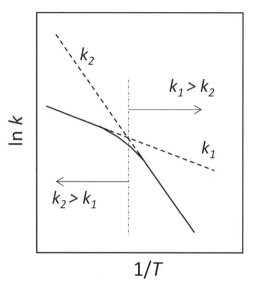

Fig. 3-2   Arrhenius plot with a concave curve upward

In the high temperature range ($k_1 \ll k_2$): $k \approx k_1 = A_1 \exp[-(E_a)_1/RT]$

Namely, the apparent activation energy will be $(E_a)_2$ in the low-temperature range and $(E_a)_1$ in the high-temperature range.

Negative temperature dependency

Some reactions have negative activation energies. In what cases are the negative activation energy possibly observed? Zhang et al. studied the carburization reaction of solid Fe with CO-CO_2 gas mixtures [2]. The weight change curves with time in the CO atmosphere at various temperatures are shown in Fig. 3-3.

In their experiments, the reaction did not proceed at less than 973 K. At more than 973 K, the carburization rates increased with the temperature up to 1273 K. At more than 1273 K, the rates decreased with the further rise in temperature. The plot of the weight gains after 3.5 ks as a function of 1/T is shown in Fig. 3-4 to clearly demonstrate this abnormal temperature dependency. The carburization rate decreases with the temperature at more than 1273 K.

What kind of reaction mechanisms make it possible to have such a negative temperature dependency?

In the previous section 2.4.4, we discussed the combination of reversible and consecutive reactions such as A $\rightleftarrows$ B $\rightarrow$ P. If A $\rightleftarrows$ B is the rate-determining equilibrium reaction, the overall reaction may have a negative temperature dependence depending on the conditions. Consider the case that the reversible reaction rates A $\rightleftarrows$ B is much faster than that of B $\rightarrow$ P $((k_F + k_B) \gg k_2)$. Under this condition, reaction A $\rightleftarrows$ B is practically under equilibrium condition. k_F and k_B are the rate constants of the forward and backward reaction of the reversible

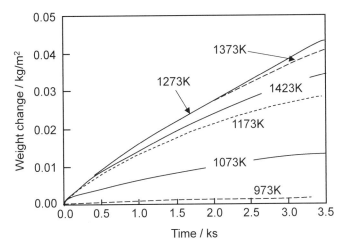

Fig. 3-3 The weight changes due to the carburization of Fe by CO at various temperatures [2].

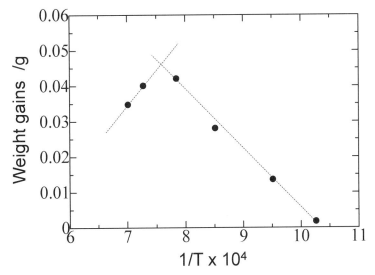

Fig. 3-4 Arrhenius-type plot of the apparent carburization reaction of Fe presented by the weight loss at 3.5 ks.

reaction A $\rightleftarrows$ B, respectively. k_2 is the rate constant of the reaction B → P.

The concentration change of the product [P(t)], which was introduced in section 2.4.4, is expressed by (3-10),

$$[P(t)] = [A(0)](1 - exp\,[-k_{eff}\,t]) \qquad \text{----- (3-10)}$$

where $k_{eff} = k_2 K_{eq}/(1 + K_{eq})$.

In the extreme case of $K_{eq} \ll 1$, k_{eff} is approximated to $k_2 K_{eq}$. K_{eq} is the equilibrium constant of the reaction A $\rightleftarrows$ B, and K_{eq} can be expressed by

$$K_{eq} = \exp(-\Delta G^0/RT) = \exp(\Delta S^0/R) \exp(-\Delta H^0/RT)$$
$$= K_0 \exp(-\Delta H^0/RT) \qquad \text{----- (3-11)}$$

where ΔG^0 is the Gibbs standard free energy change of the reaction of A $\rightleftarrows$ B, $K_0 (= \Delta S^0/R)$ is the constant fixed with temperature, and T is the absolute temperature.

The reaction constant k_2 is generally expressed by

$$k_2 = A_2 \exp(-E_2/RT) \qquad \text{----- (3-12)}$$

where A_2 is the pre-exponential term, and E_2 is the activation energy of the reaction of B $\rightarrow$ P.

Then, the k_{eff} is expressed by,

$$k_{eff} = k_2 K_{eq} = A_2 \exp(-E_2/RT) \cdot K_0 \exp(-\Delta H^0/RT) = A_2 K_0 \exp[-(E_2 + \Delta H^0)/RT]$$
$$= A_2 K_0 \exp(E_A/RT) \qquad \text{----- (3-13)}$$

Thus, under the condition of $K_{eq} \ll 1$, as described in (3-13), the apparent activation energy E_A of the overall reaction is expressed by the summation of the activation energy E_2 and the standard enthalpy ΔH^0 of the reaction of A $\rightleftarrows$ B.

$$E_A = E_2 + |\Delta H^0| \qquad \text{----- (3-14)}$$

$-$ sign and $+$ sign correspond to an exothermic reaction and endothermic reaction, respectively. When the equilibrium reaction is endothermic, $E_A = E_2 + |\Delta H^0|$, the apparent activation energy E_A is always positive, and the rate increases with temperature. When the equilibrium reaction is exothermic, $E_A = E_2 - |\Delta H^0|$, and E_A becomes smaller than E_2. In the extreme case of $E_2 < \Delta H^0$, E_A becomes negative, and the rate decreases with temperature. Thus, in the reaction of A $\rightleftarrows$ B $\rightarrow$ P, if the activation energy of the reaction A $\rightleftarrows$ B is much larger than that of B $\rightarrow$ P, the apparent activation energy of the overall reaction of A $\rightleftarrows$ B $\rightarrow$ P has a negative temperature dependency. The carburization of Fe by CO is apparently expressed by

$$2CO \rightarrow C + CO_2 \qquad \text{----- (3-15)}$$

The reaction (3-15) can be separated into the following reactions.

$$CO = C + O \qquad \text{----- (3-16)}$$
$$O + CO \rightarrow CO_2 \qquad \text{----- (3-17)}$$

The reaction (3-16) is known to be very fast and is equilibrated with general experimental conditions. Thus, A $\rightleftarrows$ B will correspond to (3-16), and B $\rightarrow$ P to (3-17). Unfortunately, no reliable values of ΔH^0 of the reaction (3-16) and E_2 of the reaction (3-17) have been reported. The apparent activation energy of (3-15) cannot be evaluated. An example of the quantitative evaluation of negative activation energy is shown in Section 7.1.

3.3 Theoretical meaning of the Arrhenius equation (Boltzmann distribution)

For the Arrhenius type equation of $k(T) = A_p \exp(-E_a/RT)$, it is important to understand the physical meaning of the pre-exponential A_p and the activation energy E_a. To promote reactions, (1) molecules must collide or meet each other, and (2) these molecules must have sufficient energy for the reaction to occur. A_p contains the collisions term of all molecules. Generally, the ratio of the molecules with a particular energy over the total energy of molecules in the reaction system is presented by the Boltzmann distribution. Then the proportion of the molecules with energies of more than E_a is expressed by $\exp(-E_a/RT)$. Consequently, the rate coefficient can be expressed by $A_p \exp(-E_a/RT)$.

The activation energy is also thought to be an energy barrier between the reactants and the products. To understand the meaning of the activation energy, the enthalpy diagram for a chemical reaction as shown in Fig. 1-2 is helpful. This is a schematic of thermodynamic energies associated with the reactants, the products, and the *hypothetical transition state* connecting them.

3.4 Effect of total pressure on the reaction rates

For the ideal gas with the constant volume V and total pressure P, the molar concentration of component i ($[i]$) and the partial pressure of i component (P_i) are related by the equation of state.

$$P_i V = n_i RT \qquad \text{----- (3-18)}$$

or

$$(n_i/V) = [i] = P_i/RT \qquad \text{----- (3-19)}$$

That is, the molar concentration of each component is linearly related to the partial pressure of each component. Within the pressure range where the gas behaves as an ideal gas, the partial pressure of each component and molar concentration increases linearly with an increase of total pressure, as shown in eq. (3-19). Thus, except under very high-pressure conditions, the effect of total pressure on reaction rates can be easily assessed by the increase in a concentration corresponding to the increase in total pressure.

3.4.1 Effect of total pressure on the activity of carbon

Under 0.1 MPa, oxygen potential pO_2 in CO-CO_2 gas mixture is fixed by the ratio pCO_2/pCO based on the following reaction:

$$CO + O_2/2 = CO_2 \qquad \text{----- (3-20)}$$
$$pO_2 = K_O(pCO_2/pCO)^{1/2} \qquad \text{----- (3-21)}$$

where K_O is the equilibrium constant of reaction (3-18). When the total pressure is increased to N MPa from 0.1 MPa, the oxygen potential is calculated at

$$pO_2 = K_O(NpCO_2/NpCO)^{1/2} = K(pCO_2/pCO)^{1/2} \qquad \text{----- (3-22)}$$

Namely, the oxygen potential in the CO-CO_2 gas mixture is independent of the total pressure. Carbon activity in CO-CO_2 gas mixture under 0.1 MPa, aC, is fixed by the ratio $(pCO)^2/pCO_2$ based on the following reaction:

$$C + CO_2 = 2CO \qquad \text{----- (3-23)}$$
$$aC = K_C(pCO)^2/pCO_2 \qquad \text{----- (3-24)}$$

where K_C is the equilibrium constant of reaction (3-23). When the total pressure is increased to N from 0.1 MPa, the aC is calculated to

$$aC(\text{at } N \text{ MPa}) = K_C(NpCO)^2/NpCO_2, = N(K_C(pCO)^2/pCO_2) = N \, aC(\text{at } 0.1 \text{ MPa atm})$$
$$\text{----- (3-25)}$$

Namely, with the increase of total pressure to N, the aC becomes N times larger. However, with the decrease of the total pressure by evacuation or diluting by inert gas in the CO-CO$_2$ gas mixture, the pO_2 does not change, but the carbon activity decreases.

The carbon activity in the CH$_4$-H$_2$ gas mixture, aC, is fixed by $pCH_4/(pH_2)^2$ based on the following reaction:

$$CH_4 = C + 2H_2 \qquad \text{----- (3-26)}$$
$$aC = K_{CH} \, pCH_4/(pH_2)^2 \qquad \text{----- (3-27)}$$

where K_{CH} is the equilibrium constant of the reaction (3-26). Thus, the aC decreases with the increase of the total pressure, and the dilution of the CH$_4$-H$_2$ gas mixture increases the carbon activity! Therefore, care must be taken when calculating the equilibrium carbon activity at the interface when the CH$_4$-H$_2$ gas mixture is diluted with Ar.

3.5   The rate equations for varying volume systems

In this section, the rate equation is introduced for the case when the volume of the system varies with reactions under constant total pressure. Details of the reacting system with varying volumes can be found elsewhere [3].

3.5.1   The relation between conversion and concentration for the varying volume systems

The general form for the reaction rate of component A in variable volume system under constant total pressure is

$$r_A = (1/V)(dN_A/dt) = (1/V)d([A]V)/dt = (1/V)(Vd[A]/dt + [A]dV/dt)$$

or

$$r_A = d[A]/dt + ([A]/V)(dV/dt) \qquad \text{------(3-28)}$$

where N_A is moles of component A, V is a volume of the reacting system with time. For a constant-volume system, the second term drops out, leaving the simple expression.

$$r_A = d[A]/dt \qquad \text{------(3-29)}$$

For a variable volume system, however, the second term in (3-28) must be evaluated to find r_A. By assuming that the volume of the reacting system varies linearly with the fractional conversion X_A, or

$$V = V_0(1 + \varepsilon_V X_A) \text{ or } X_A = (V - V_0)/V_0\varepsilon_V \qquad \text{----- (3-30)}$$

where V_0 is the initial volume of the reacting system, V is the volume at time t, X_A is the fractional conversion of A at time t, respectively. ε_V is the fractional change in volume of the reaction system, and defined by,

$$\varepsilon_V = [V(\text{at } X_A = 1) - V(\text{at } X_A = 0)]/V(\text{at } X_A = 0) \qquad \text{----- (3-31)}$$

In most gas-phase homogeneous reaction systems, this linearity relation (3-31) is endorsed.

As an example of the use of ε_V, consider the carbon solution-loss reaction

$$CO_2 + C \rightarrow 2CO \qquad ----- (3\text{-}32)$$

By starting with pure CO_2, and from the stoichiometry of the reaction (3-32),

$$\varepsilon_V = (2 - 1)/1 = 1$$

When CO_2 is diluted with an inert gas such as 50% of Ar,

$$CO_2 + C + Ar \rightarrow 2CO + Ar \qquad ----- (3\text{-}33)$$

$$\varepsilon_V = (2 + 1 - (1 + 1))/(1 + 1) = 0.5$$

Namely, ε_V must account for not only the reaction stoichiometry but also the presence of inert gas.

Using (3-30), the concentration [A] can be expressed by using X_A,

$$[A(t)] = N_A/V = N_{A0}(1 - X_A)/V_0(1 + \varepsilon_V X_A)$$
$$= [A(0)](1 - X_A)/(1 + \varepsilon_V X_A) \qquad ----- (3\text{-}34)$$
$$[A(0)] = N_{A0}/V_0.$$

Thus,

$$[A(t)]/[A(0)] = (1 - X_A)/(1 + \varepsilon_V X_A) \qquad ----- (3\text{-}35)$$

The relationship between $[A(t)]/[A(0)]$ and X_A is shown in Fig. 3-5, with several values of ε_V. When the volume stays constant during a reaction, the relation between $[A(t)]/[A(0)]$ and X_A is linear. However, the relation between $[A(t)]/[A(0)]$ and X_A deviates from the linearity with increases of ε_V.

When the concentration is halved, X_A becomes 0.5 with the case of $\varepsilon_V = 0$. However, in the case of $\varepsilon_V = 3$, the conversion ratio was only 0.2, although the concentration was reduced to 50%. This difference is because the concentration rapidly decreased due to the increase in

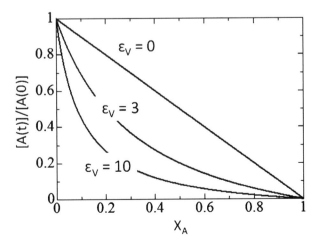

Fig. 3-5 $[A(t)]/[A(0)]$ as a function of X_A with various ε_V.

volume as the reaction proceeded. In other words, the reaction rate rapidly decreases with the progress of the reaction compared with that of the constant volume reaction. The practical approach to reducing the volume change effect is diluting the reacting gas by inert gas since the ε_V becomes smaller with the increase of the inert gas ratio, as demonstrated by the reaction (3-30).

3.5.2 Rate equation of first-order irreversible reaction with varying volumes

The general rate equation with varying volume reacting systems is expressed by

$$-r_A = -(1/V)(dN_A/dt) \quad\quad\quad ----- (3\text{-}36)$$

Since $N_A = N_{A0}(1\text{-}X_A)$, and replacing V from (3-30)

$$-r_A = (A[0]/(1 + \varepsilon_V X_A))(dX_A/dt) \quad\quad ----- (3\text{-}37)$$

Whether the reacting volume is constant or not, the rate for a first-order irreversible reaction, is expressed by

$$-r_A = k[A]$$

where k is a rate constant.

Using (3-34),

$$-r_A = k[A] = kA[0](1 - X_A)/(1 + \varepsilon_V X_A) \quad\quad ----- (3\text{-}38)$$

Comparing (3-37) and (3-38),

$$(A[0]/(1 + \varepsilon_V X_A))(dX_A/dt) = kA[0](1 - X_A)/(1 + \varepsilon_V X_A) \quad ----- (3\text{-}39)$$

or

$$dX_A/dt = k(1 - X_A) \quad\quad\quad ----- (3\text{-}40)$$

Separating and integrating (3-40)

$$- \ln(1 - X_A) = k \quad\quad\quad ----- (3\text{-}41)$$

The rate equation with concentration in the integrated form of the irreversible first-order reaction for constant volume system is already presented in chapter 2 by

$$- \ln([A(t)]/[A(0)]) = - \ln(1 - X_A) = kt, \quad\quad ----- (2\text{-}11)$$

Compared with (2-7) and (3-41), the rate equations using fractional conversion for the irreversible first-order reaction expressed with constant volume and varying volume reacting systems have the same expression.

Further reading

O. Levenspiel: Chemical Reaction Engineering 3rd ed., John Wily & Sons, Hoboken, 1999.

M. J. Pilling and P. W. Seakins, Reaction Kinetics, Oxford Uni. Press, Oxford, 2005.

References

[1] J. Laidler: The development of the Arrhenius equation, J. Chem. Edu., 61 (1984), 494-498.

[2] X. Zhang, R. Takahashi, T. Akiyama and J. Yagi.: Tetsu to Hagane, 83 (1997), 299-304.

[3] O. Levenspiel: Chemical Reaction Engineering 3rd ed., John Wily & Sons, Hoboken, 1999,

67-69.

Chapter 4
Experimental techniques for measuring reaction rates in heterogeneous reactions

Chapter 4   Experimental techniques for measuring reaction rates

Kinetics studies which start from the measurement of reaction rates as reaction kinetics is an experimental science. Many various experimental methods have been developed to measure reaction rates. The rates are simply evaluated by measuring the change of the concentrations of reactants or products per unit time under certain conditions. When the concentrations of reactants or products are related to other physical properties, such as weight, total pressure, gas composition, thermal or electrical conductivity, the rates can be measured based on these changes if the relations between them are known. For the first-order reaction, the rate can be directly evaluated from the changes of these physical properties.

These general methods are readily applied for the measurements of homogenous reaction rates. For heterogeneous reactions, however, there are three notable concerns:

(1) the reacting surface area,

(2) the surface chemical properties such as Fe^{3+}/Fe^{2+} ratio since these may change along with the reaction. In chapter 1, it was stated that the liquid surface has essentially only one unique structure, while the solid surface structures of polycrystalline have almost infinite variations. However, even at liquid surfaces, the chemical properties sometimes change,

(3) the mass transport process to/from the surface.

Thus, specific experimental techniques are required for taking measurements of the heterogeneous reaction rate.

4.1   Change of the reacting surface area with time

The actual reaction area changes with time in many heterogeneous reactions, although the apparent geometrical surface area does not change. As an example, let's consider the following liquid Fe oxidation reaction with CO_2-CO gas mixture,

$$CO_2 \text{ (gas)} \rightarrow CO \text{ (gas)} + O(ad) \qquad \text{----- (4-1)}$$

$$O(ad) = \underline{O} \qquad \text{----- (4-2)}$$

where O(ad) and $\underline{O}$ refer to the adsorbed oxygen atoms at the liquid Fe surface and the dissolved oxygen atoms in the liquid Fe, respectively. These reactions consist of several elementary steps, but just for simplicity, only the reactions (4-1) and (4-2) are presented here to demonstrate the surface area change during the reaction. With the progress of the reactions, the concentrations of O(ad) and $\underline{O}$ gradually increase until they reach the equilibrium values that are fixed by the CO_2/CO ratio in the introducing CO-CO_2 gas mixture. It is well known that O(ad) will block the reaction sites for CO_2 gas adsorption, and the reaction occurs only at the unblocked bare Fe surface. Consequently, the apparent reaction rate gradually decreases with the increase of O(ad) since the available reaction area decreases with time. Therefore, the CO_2 dissociation rates may be expressed by

$$dpCO_2/dt = -k \, pCO_2 \, (1 - \theta(t)) \qquad \text{----- (4-3)}$$

where $\theta(t)$ is the surface fraction of O(ad) with time. When the surface is fully covered by

O(ad), $\theta(t) = 1$, and reactions will not occur. To calculate the reaction rate coefficient k from the measurements of the apparent reaction rate ($dpCO_2/dt$), we need to know the surface area or $\theta(t)$ as a function of time.

4.2 Change of the surface chemical properties with time

To understand the surface property changes during the reaction, let's consider the molten iron oxide reduction reaction with CO gas. There are two types of molten iron oxide reduction reactions: (1) reduction without the formation of metallic Fe, and (2) reduction with metallic Fe formation. The reduction without Fe formation can be expressed by

$$CO + O^{2-} \rightarrow CO_2 + 2e^- \qquad \text{----- (4-4)}$$
$$Fe^{3+} + e^- \rightarrow Fe^{2+} \qquad \text{----- (4-5)}$$

where 'e$^-$' refers to the electron. In this reduction, along with the oxygen (O^{2-}) removal from the liquid iron oxide surface, the ratio Fe^{3+}/Fe^{2+} at the surface decreases to keep the electrical neutrality until Fe finally appears. The Fe^{3+}/Fe^{2+} ratio change influences the overall molten iron oxide reduction rate. In other words, to evaluate the reaction rate coefficient from the apparent reaction rate measurement, the effect of the ratio Fe^{3+}/Fe^{2+} must be considered. Once Fe is formed, the Fe^{3+}/Fe^{2+} ratio does not change, and the reaction rates can simply be determined by pCO.

Thus, the reaction rates of heterogeneous reactions must be measured under the constant reaction area as well as the unchanged chemical properties (such as Fe^{3+}/Fe^{2+}) at the surface. If not, the change of the reaction area and chemical properties with time must be known.

4.3 How to maintain the reacting area constant and keep the surface properties unchanged

4.3.1 Isotopic exchange reactions

One of the most useful and effective methods to measure chemical reaction rates while maintaining the reaction area constant and surface chemical properties unchanged during the reaction is the "Isotopic exchange reaction" technique. The reaction area and chemical properties do not change during an isotope exchange reaction since the reactions are carried out **under equilibrium conditions**. Even under equilibrium conditions, kinetics information such as reaction rates is possibly obtained using isotope exchange reactions.

An isotopic exchange reaction involves substituting one isotope of an element with another isotope of the element in the molecules of a given substance without changing their compositions. For example, if ^{14}C enriched CO_2 is mixed with CO, and this gas mixture is introduced onto metal or oxide surfaces, then the following oxygen exchange reaction takes place:

$$^{14}CO_2 + {}^{12}CO \rightarrow {}^{14}CO + {}^{12}CO_2 \qquad \text{----- (4-6)}$$

By this exchange reaction, the CO gas is enriched by the heavy isotope ^{14}C, and the CO_2 becomes depleted by the heavy isotope ^{14}C.

The reaction (4-6) occurs by the following two partial reactions:

$$^{14}CO_2 \rightarrow {}^{14}CO + O(ad) \qquad\qquad ----- (4\text{-}7)$$
$$^{12}CO + O(ad) \rightarrow {}^{12}CO_2 \qquad\qquad ----- (4\text{-}8)$$

Thus, the measurement of the formation rate of ^{14}CO gives the rate of $^{14}CO_2$ dissociation reaction (4-8) on the metal or oxide surfaces corresponded with a particular experimental condition, provided that homogeneous gas reaction and reaction at the surfaces of apparatus system can be neglected. The isotope effect, which results from the difference in mass of ^{12}C and radioactive ^{14}C is negligible since the chemical properties for isotopes of the same elements are almost identical, and the relative differences in their atomic masses are not significant. Therefore, the dissociation reaction behavior of $^{14}CO_2$ can be practically the same as that of $^{12}CO_2$. Consequently, the formation rate of ^{14}CO can give the rate of the untagged reaction (4-9):

$$CO_2 \rightarrow CO + O(ad) \qquad ----- (4\text{-}9)$$
$$CO + O(ad) \rightarrow CO_2 \qquad ----- (4\text{-}10)$$

The rates of chemical reactions are conventionally measured under conditions where a driving force has been established by introducing concentration or chemical potential differences in the system. However, with isotope exchange reactions, the rates of reactions are possibly measured even when there is no apparent driving force or under equilibrium conditions.

Why can we measure the reaction rate even under the equilibrium condition using isotope exchange reactions?

Let's think about the reaction of molten iron with the CO_2-CO gas mixture again. Assume a case in which, initially, CO_2 and CO do not contain the ^{14}C isotope. Under equilibrium conditions, the concentrations of CO_2 and CO in the inlet and outlet gas mixture are the same. However, the reactions (4-9) and (4-10) always co-occur, even under equilibrium conditions. In other words, the decrease amount of CO_2 by the reaction (4-9) is precisely compensated by the increasing amount of CO_2 by the reaction (4-10). Thus, the amounts or concentrations of CO_2 and CO are unchanged. In this case, it is impossible to measure the CO_2 dissociation and formation rate by the usual methods.

Suppose the liquid Fe is equilibrated with CO_2-CO gas mixture. Then, a small amount of CO_2 is replaced by the same amount of ^{14}C labeled $^{14}CO_2$. The chemical properties of $^{14}CO_2$ and ^{14}CO are practically the same as these of $^{12}CO_2$ and ^{12}CO, respectively. Thus, all values, including the contents of $\underline{O}$ and O(ad) in the molten Fe after introducing $^{14}CO_2$ contained gas mixture, are the same as that before$^{14}CO_2$ addition, or nothing seems to be changed after introducing $^{14}CO_2$.

Under the apparent equilibrium condition, the amounts of CO_2 (= $^{14}CO_2$ + $^{12}CO_2$) and CO (= ^{14}CO + ^{12}CO) in the inlet and outlet do not change. However, the reaction (4-7), (4-8), (4-9), and (4-10) always occur. In this isotope exchange reaction, $^{14}CO_2$ decreases, and ^{14}CO increases in the outlet gas due to the $^{14}CO_2$ dissociation reaction of (4-7). From the measurement of the increasing amount of ^{14}CO per unit time in the outlet gas, the $^4CO_2$ dissociation rate can be calculated.

Since $^{14}CO_2$ dissociation rate is measured based on the amount of ^{14}CO is formed by the reaction (4-7), the backward reaction of (4-7)' must be avoided to evaluate the true chemical reaction rate,

$$^{14}CO + O(ad) \rightarrow {}^{14}CO_2 \qquad \text{----- (4-7)'}$$

The backward diffusion of formed ^{14}CO in the leaving gas from the interface should be avoided and is easily achieved by supplying the reasonably high flow rate of the gas mixture.

Experiments by isotope exchange reactions are quite straightforward. $^{14}CO_2$ contained CO_2-CO gas mixture is introduced to the sample surface and selectively separate CO_2 from the reacted gas mixture with liquid nitrogen or molecular sieve followed by measurements of radioactivity of ^{14}CO. The separated ^{14}CO contained CO gas is sometimes oxidized to CO_2 for easy measurements of $^{14}CO_2$ amounts. The amount of ^{14}CO is measured by Q-mass spectroscopy or Geiger-Müller counters.

Depending on the characteristics of the exchange reaction systems, several equations such as the McKay equation to evaluate the reaction constants in the exchange reaction are applied. For the exchange reaction of (4-6), the following equation is used to calculate the reaction rate constant k of the reaction (4-7).

$$k = \frac{V_t}{ART} \frac{1}{1+B} ln \left[\frac{1}{1 - P^{14}CO/(P^{14}CO)_{eq}} \right] \qquad \text{----- (4-11)}$$

where V_t is the total volume flow rate of the gas mixture, A is the surface area of the sample, B is the CO_2/CO ratio, and $p^{14}CO$ and $(p^{14}CO)_{eq}$ are the partial pressure of tagged CO after the reaction and if complete isotope equilibrium were to be reached. Details for deducing eq. (4-11) can be found elsewhere [1].

4.3.2 Effective use of the saturated conditions

For the reactions under saturated conditions, the concentration of a saturated element or component is fixed. Take molten iron oxide reduction by CO gas as an example. The Fe^{3+}/Fe^{2+} ratio in molten iron oxide influences the molten iron oxide reduction rate. In other words, to evaluate the reaction rate coefficient, the effect of the ratio Fe^{3+}/Fe^{2+} must be considered. Nagasaka et al. studied molten iron oxide reduction by CO gas with Fe saturated conditions [2]. In their experiments, molten iron oxide was held in a Fe crucible. The experimental system used is shown in Fig. 4-1. The reduction rate was evaluated by the weight decrease of molten iron oxide using a thermal balance. In their experiments, before starting experiments, molten iron oxide held in the Fe crucible is maintained at a particular temperature with Ar for about 1 hr to equilibrate molten iron oxide with Fe. Thus, iron oxide is supposed to be Fe saturated.

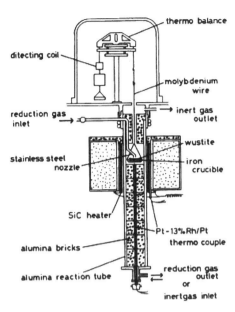

Fig. 4-1 Thermogravimetric experimental system by Nagasaka et al [2].

In the Fe saturated condition, the Fe^{3+}/Fe^{2+} ratio in molten iron oxide does not change during the reduction process until all Fe^{2+} is reduced to metallic Fe.

The molten iron oxide reduction by CO gas under Fe saturated condition is expressed by

$$CO + O^{2-} \rightarrow CO_2 + 2e^- \qquad ----- (4\text{-}12)$$
$$Fe^{2+} + 2e^- \rightarrow Fe \qquad ----- (4\text{-}13)$$

where 'e$^-$' refers to an electron. The electron conductivity in molten iron oxide is significantly rapid so that the reaction of (4-12) and (4-13) can proceed at different positions. Consequently, the reaction (4-12) or O^{2-} removal reaction proceeds with constant Fe^{3+}/Fe^{2+} ratio.

As another example: Decarburization of C saturated molten Fe by CO_2 gas.

Sain and Belton carried out the decarburization reaction of carbon saturated molten Fe by CO_2 gas [3]. The used experimental system is shown in Fig. 4-2.

Liquid Fe was held in an Al_2O_3 crucible with graphite at the bottom of the crucible with the vigorous stirring of liquid Fe by high-frequency induction, and reacting CO_2 gas was injected onto the surface of molten Fe as shown in Fig. 4-2. The decarburization reaction is expressed by

$$\underline{C} + CO_2 \rightarrow 2CO \qquad ----- (4\text{-}14)$$

Without the graphite disk at the bottom, the carbon content in the molten Fe gradually decreases as the reaction proceeds (4-14). Then, the dissolved oxygen $\underline{O}$ content in Fe melt increases. With an increase of $\underline{O}$ content, the adsorbed oxygen O_{ad} at the surface increase, blocking the reacting sites. With the proposed design of the reacting system, carbon is continuously removed by the reaction (4-14). However, carbon is also always supplied from

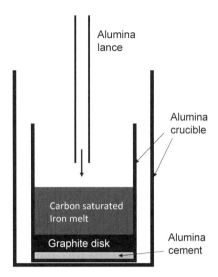

Fig. 4-2 Gas jetting experimental system by Sain et al. [3].

the graphite at the bottom by the vigorous stirring of the Fe melt, so it can be assumed that the molten Fe is always saturated with carbon even during the reaction. Under this carbon-saturated condition, the decarburization reaction of the Fe-C melt can be carried out with the constant carbon content corresponding to the carbon-saturated condition. Under the carbon-saturated condition, the amount of dissolved oxygen content is negligibly small, and therefore the amount of the site-blocking O_{ad} also can be very small. In other words, we can reasonably assume that any reaction sites at the molten Fe surface are not blocked at all. Thus, the decarburization rate at the carbon saturated Fe-C melt corresponds to the rate at the bare surface with no blocking of the reacting site.

4.4   Effect of mass transfer on the reaction rates

In heterogeneous reactions, reactants must move to the surface, and products must leave from the surface. Sometimes, products remain near the interface for some time. Thus, to a greater or lesser extent, mass transfer resistance always exists in heterogeneous reactions. The effect of the mass transfer resistance on the overall apparent reaction rate must be eliminated or assessed to evaluate accurate chemical reaction rates from kinetics experiments.

In a simple gas-solid or gas-liquid reaction experiment, a popular conventional technique to eliminate the effect of mass transfer is to carry out experiments with varying gas flow rates and find out the 'plateau region', in which the reaction rate does not change with the gas flow rates. This situation is shown in Fig. 4-3.

Because mass transfer coefficients depend on gas velocity, the existence of the plateau region may suggest that the mass transfer coefficients in the plateau region are so large that the mass transfer resistance term in the overall rate is negligible.

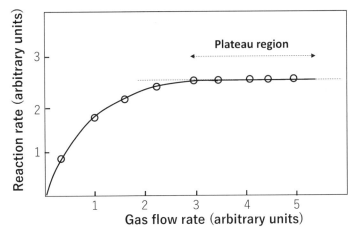

Fig. 4-3 The apparent reaction rates with gas flow rates

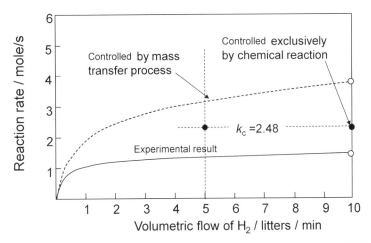

Fig. 4-4 Variation of the initial reaction rate of 1 cm diameter hematite spheres at 1073 K with the volumetric flow rate of pure hydrogen. Dashed lines are for a reaction exclusively controlled by mass transfer [4].

However, A. W. D. Hills confirmed that this assumption was wrong, and convective mass transfer still played even in the plateau region based on his careful experiments of Fe_2O_3 reduction with varying H_2 gas flow rates [4]. His reduction results of 1.0 cm diameter hematite sphere are presented in Fig. 4-4. The measured reaction rate showed the plateau under the high flow rate of more than about 5 l/min. Suppose the resistance of the mass transfer process is negligible. In that case, the chemical reaction rate constant k_{app} can be calculated from the measured rate in the plateau region by using the following equation.

$$\text{Apparent rate} = k_{ap} \, pH_2^b \qquad \text{----- (4-15)}$$

where pH_2^b is the inlet H_2 partial pressure. In the figure, the calculated reduction rate exclusively controlled by mass transfer (indicated by dashed lines) is also shown. It has only about 3 times larger values even under the flow rate of 10 l/min.

It means that the mass transfer resistance cannot be neglected even under the plateau region in his experiments, or the reaction is supposed to be under the mixed-controlled regime. Then, we must consider the two processes, gas film and the chemical reaction at the interface, to describe the reaction rate [5]. In other words, the H_2 partial pressure at the interface is not pH_2^b but pH_2^i. pH_2^b and pH_2^i are the inlet H_2 partial pressure and H_2 partial pressure at the interface. The rate of the transfer of H_2 from gas to the hematite surface given by the rate expression,

$$Rate = m(pH_2^b - pH_2^i) \qquad \text{----- (4-16)}$$

where m is the mass transfer coefficient. The rate of the interfacial reaction rate is given by

$$Rate = k_c pH_2^i \qquad \text{----- (4-17)}$$

where k_c is the true chemical reaction rate coefficient.

Since the mass transfer rate and the chemical reaction rate are equal,

$$m(pH_2^b - pH_2^i) = k_c pH_2^i \qquad \text{----- (4-18)}$$

From eq. (4-18),

$$pH_2^i = (m/(m + k_c)) pH_2^b \qquad \text{----- (4-19)}$$

Since the measured apparent reaction rate (corresponds to the solid line in Fig. 4-4 is also equal to the interfacial reaction rate under the mixed-controlled regime,

$$Apparent\ rate = k_{ap} pH_2^b = Rate = k_c pH_2^i = (m/(m + k_c)) pH_2^b \qquad \text{----- (4-20)}$$

where k_{ap} is the apparent reaction coefficient, and is expressed by

$$k_{ap} = (k_c m/(m + k_c)), \text{ or}$$

$$1/k_{ap} = 1/k_c + 1/m \qquad \text{----- (4-21)}$$

The true chemical reaction rate (without the mass transfer resistance) with flow rates of 5 and 10 l/min is calculated at about 2.48 mole/s for both flowrates using the eq. (4-21), and are presented by closed circles in Fig. 4-4. The estimated true chemical reaction rate (a closed circle) is about 1.6 times larger than the experimental results. Thus, it is incorrect to conclude that convective mass transfer can be ignored when the constant reaction rates are observed in the plateau region.

Units of mass transfer coefficient

Mass transfer coefficients are not physical properties like the diffusion coefficient. They differ from case to case, depending on their definition. In mass transfer, various concentrations are conventionally used, such as mass fraction, mole fraction, partial pressure, and mole concentration. As a result, different units of mass transfer coefficients are often used even for the same mass transfer flux.

Mass flux N_A [mole/cm^{2}s] can be expressed by using mole concentration C [mole/cm^3] difference and mass transfer coefficient m,

$$N_A = m\Delta C \qquad \text{----- (4-22)}$$

ΔC is the appropriate concentration difference and, most commonly, that of the interface concentration and the equilibrium or asymptotic concentration. In this case, the unit of m is [cm/sec]. By applying the general gas equation of (4-23),

$$PV = nRT \qquad \text{----- (4-23)}$$

concentrations can be expressed by partial pressure P.

$$C = n/V = P/RT \qquad \text{----- (4-24)}$$

Combining (4-22) and (4-24),

$$N_A = m\Delta(P/RT) = (m/RT)\,\Delta P = m_p \Delta P \qquad \text{----- (4-25)}$$

where m_p is the mass transfer coefficient when using the partial pressure difference. The unit of m_p is [mol/(cm$_2$ s atm)]. This unit often appears as a chemical reaction rate constant.

Through many experiments, it is often found that the mass transport coefficient *m* in fluid-solid reactions is proportional to the square root of flow rate *V*;

$$m \propto (V)^{1/2} \qquad \text{----- (4-26)}$$

Since *m* is proportional to the square root of the flow rate, the increment of *m* becomes smaller with the flow rates. Under a high flow rate condition, *m* only slightly increases and appears almost constant within experimental errors. Namely, often observed almost constant rates in the 'plateau region' may still contain non-negligible amounts of mass transfer resistance. In other words, it is difficult to eliminate mass transfer resistance in kinetics experiments without special ingenuity. Thus, to obtain an apparently constant reaction rate in the plateau region, we must evaluate the effect of mass transfer resistance on the measured reaction rate. One approach is to calculate the mass transfer coefficient for the applied experimental condition based on mass transfer correlations. Once the mass transfer coefficient is evaluated, the true chemical reaction rate without mass transfer resistance can be evaluated.

Decreasing temperature due to high flow rate.

The high gas flow rate reduces the mass transport resistance as well as the heat transport resistance. As a result, the surface temperature of the sample may decrease under a high flow rate, especially for an external heating system, resulting in a decrease in the reaction rate. This decrease in the reaction rate may introduce the plateau region. Therefore, it is necessary to pay attention to the decrease in the surface temperature in the external heating experimental system.

4.5 How to evaluate the mass transfer coefficient?

Mass transfer coefficients are not physical properties like diffusion coefficients. They differ from case to case depending on experimental conditions. With the help of experimental observations, correlations for mass transfer coefficients have been developed for standard cases

Table 4-1 Dimensionless numbers for mass transfer

Group	Physical meaning
Sherwood number Sh, mL/D	mass transfer velocity/diffusion velocity
Stanton number St, m/v_0	mass transfer velocity/flow velocity
Schmidt number Sc, v/D	diffusivity of momentum/diffusivity of mass
Lewis number Le, α/D	diffusivity of energy/diffusion of mass
Prandtl number Pr, v/α	diffusivity of momentum/diffusivity of energy
Reynolds number Re, uL/v	inertia forces/viscosity forces

k: mass transfer coefficient, L: characteristic length, D: diffusivity, u: flow speed, v: kinematic viscosity, α: thermal diffusivity.

(e.g. fluid flow through a packed bed of particles, gas bubbles rising in a tank, flow over surfaces). These coefficients and are rarely reported as individual values but as correlations of dimensionless numbers.

The characteristics of the common dimensionless groups frequently used in mass transfer correlations are given in Table 4-1. The mass transfer coefficient is most often presented as a Sherwood number, Sh. A Schmidt number, Sc, generally reflects the effect of the diffusion coefficient in the mass transfer process. The effect of flow rate is evaluated by using a Reynolds number, Re, for forced convection.

With the help of experimental observations, many correlations for mass transfer coefficients have been developed for standard cases (e.g., gas flow over a solid sphere, fluid flow through a packed bed of particles, gas bubbles rising in a tank, and falling films, etc.).

4.6 Chapman-Enskog theory

The dimensionless numbers consist of physical properties, such as viscosity and diffusivity. These properties vary with experimental conditions, such as gas temperature and gas species. Thus, we must know these values for particular gas composition and temperature to calculate the mass transfer coefficient. The Chapman and Enskog theory provides estimation methods to meet these demands for various transport coefficients such as viscosity, diffusivity, and thermal conductivity. The theory was introduced by Chapman and Enskog independently in 1916 and 1917. [6]

The coefficient of viscosity at absolute temperature T of molecular weight of M is expressed in terms of parameters σ and Ω_μ as,

$$\mu = 2.6693 \times 10^{-5} (MT)^{1/2}/\sigma^2\Omega\mu \qquad \text{---- (4-27)}$$

The mass diffusivity D_{AB} for a binary gas system A-B of molecular weight of M_A and M_B at absolute temperature T under the total pressure of p (atm) is expressed in terms of parameters σ_{AB} and Ω_{AB} as,

$$D_{AB} = 0.0018583 (T^3 (1/M_A + 1/M_B)^{1/2}/p\sigma_{AB}{}^2\Omega AB \qquad \text{----- (4-28)}$$

Details of these equations and the parameters can be found elsewhere.

4.7 Frequently used mass transfer correlations

4.7.1 Ranz-Marshall correlation

Mass transfer for flow past a single sphere has been the subject of many studies. Since the pioneering studies of Frössling and Vyrubov, several attempts to develop correlations, including Ranz and Marshall [7], have been widely quoted and used. It is expressed by

$$Sh = 2 + 0.6Re^{1/2}Sc^{1/3}, \qquad 0 \leq Re \leq 200, 0 \leq Sc \leq 250 \quad \text{----- (4-29)}$$

Sherwood number Sh is defined by $Sh = mL/D$, where m is mass transfer coefficient, L is the characteristic length such as a particle's diameter, and D is the diffusion coefficient.

Based on eq. (4-29), Sh will change by the flow rate and the characteristic length L. The flow rate (an increase of Re) enhances mass transfer for flow past a single sphere. From the definition of Sh, the mass transfer coefficient m can be expressed by

$$m = (ShD)/L \qquad \text{----- (4-30)}$$

From Eq. (4-30), mass transfer coefficient m increases with a decrease of the characteristic length L, since Sh is constant once the flow rate is fixed. Thus, by using fine particles, *the* mass transfer is significantly enhanced. Generally, the gas diffusion coefficients for most of the gas mixture are 0.1-0.25 cm^2/s and almost independent of temperature.

Tsukihashi *et al.* carried out fine iron oxide reduction with CO gas in a gas conveying system [8]. The reacting system is shown in Fig. 4-5.

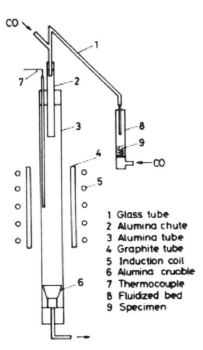

1 Glass tube
2 Alumina chute
3 Alumina tube
4 Graphite tube
5 Induction coil
6 Alumina crucble
7 Thermocouple
8 Fluidized bed
9 Specimen

Fig. 4-5 Experimental apparatus for fine iron oxide reduction with CO gas in a gas conveying system [8].

In this system, dispersedly suspended iron oxide particles in gas flow are introduced into the reactor. Since the particles do not interact and behave as a single particle in the flow, the Ranz-Marshall correlation can be applied for the mass transfer around each particle. In this system, the particles are conveyed by gas flow, and their moving velocity is practically the same as the gas flow velocity. Thus, the flow velocity around the particles can be assumed to be almost zero. Then, it can be assumed that $Re \approx 0$ and $Sh \approx 2.0$. Then,

$$m = 2.0D/L \qquad ----- (4\text{-}31)$$

The average diameter of the fine iron oxide used was about 100 µm. By applying (4-21) for their experimental conditions, m was calculated to be about 2900 cm/sec, and the measured apparent chemical reaction rate coefficient was about 20 cm/sec. From these values, the effect of mass transfer on the chemical reaction rates can be neglected under their experimental conditions. That is, the measured reaction rates reflect true chemical reaction rates. Entrained gas flow reacting system can be an attractive method to conquer mass transfer resistance.

4.7.2   Taniguchi mass transfer correlation

Various mass transfer correlations have been proposed for the case of the impinging jet onto the surface of a melt in a crucible. The typical geometrical setup of a gas imping system is shown in Fig. 4-6 [9].

This type of experimental setup is extensively used, especially in the study of high-temperature kinetics. However, despite its importance, only a few studies on the mass transfer

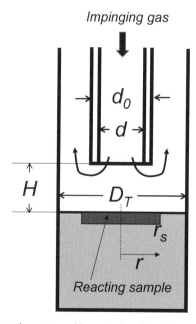

Fig. 4-6   The reaction system with gas impinged onto the sample surface [9].

correlations corresponding to these experiments have been reported. Taniguchi et al. have re-examined this system by studying the sublimation of naphthalene and the evaporation of pure liquids into the nitrogen stream [9]. They obtained the following correlation

$$Sh = 0.40(\pm 0.13)(r_s/d)^{-1} Re^{0.66} Sc^{0.5} \qquad ----- (4\text{-}32)$$

For $(H/d) \leq (H/d)_c$

where $(H/d)_c = 0.0046\ Re^{0.68}(r_s/d)^{1.5}\ \exp\ [3.98(d_0/D_T)][(\exp Sc)/Sc]$

and the additional symbols are

d: inside diameter of gas imping tube

d_0: outside diameter of gas imping tube

H: the distance between the nozzle and the sample surface

r_s: radius of reacting surface (crucible)

D_T: inner diameter of the crucible.

These symbols correspond to those in Fig. 4.6.

Belton et al. carried out the desulphurization reaction of molten Fe-S alloy with H_2. [10] It was found that the S removal rates strongly depended on the flow rates. To evaluate the effect of the mass transfer resistance, they applied their results to the mass transfer correlation by Taniguchi *et al.* [9]. As shown in Fig. 4-7, the derived Sherwood numbers fit smoothly for

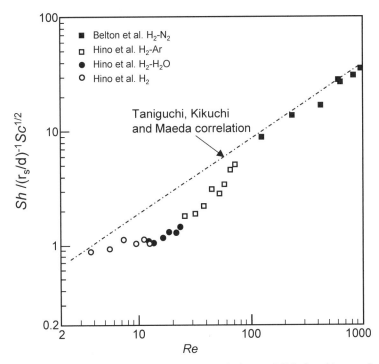

Fig. 4-7 Desulphurization results by Belton *et al.* [10] and Hino *et al.* [11] plotted in accordance with correlation (4-32).

these results at high Re numbers. That is, the desulphurization reaction was fully controlled by the mass transfer process. The deviation under Re of 70 is due to the natural convection effect. The good agreement indicates that the mass transfer process fully controlled the S removal rates in their experiments.

Further reading

R. B. Bird, W. E. Stewart, and E. N. Lightfoot: Transport Phenomena, Joh, Wiley & Sons, 1960.

G. H. Geiger: Transport Phenomena in Metallurgy, Addison-Wesley, 1973

References

[1] A. Ozaki: Isotope Studies of Heterogeneous Catalysis, Kodansha Ltd., Tokyo and Academic Press Inc., New York, NY, 1977, 25-27.

[2] T. Nagasaka, Y. Iguchi and S. Ban-ya: Testu to Hagane, 71 (1985), 204-211.

[3] Sain and Belton: Metall. Trans. B, 7B, (1976), 235-244.

[4] W. D. Hills: Metall. Trans. B, 9B, (1978), 121-128.

[5] O. Levenspiel: Chemical Reaction Engineering 3rd ed., John Wily & Sons, Hoboken, 1999, 369-74.

[6] R. B. Bird, W. E. Stewart, and E. N. Lightfoot: Transport Phenomena, Joh, Wiley & Sons, 1960. 22-23, 235, 510-513.

[7] W. E. Ranz and W. R. Marshall: Chem. Eng. Prog., 48, (1952), 141-146, 173-180.

[8] F. Tsukihashi, K. Kato, K. Otsuka and T. Soma: Trans. ISIJ, 22 (1982), 688-95.

[9] Taniguchi, A. Kikuchi, and S. Maeda: Tetsu to Hagané, 62 (1976), 191-200.

[10] G. R. Belton and R. A. Belton: Trans ISIJ, 20 (1980), 87-91.

[11] M. Hino, S. Ban-ya and T. Fuwa: Tetsu to Hagané, 62 (1976), 33-42.

Chapter 5
Adsorption phenomena

Chapter 5　Adsorption phenomena

An early and important discovery in the history of heterogeneous reactions was the observation by Faraday, who found that H_2 and O_2 reacted at Pt surface in 1834. This phenomenon means that molecules must first become attached to a surface before reacting. Then Langmuir carried out the quantitative studies of the oxidation reaction kinetics of CO and H_2 on Pt in 1921. Based on these reaction kinetics results, he established a well-known Langmuir adsorption isotherm. Langmuir won the Nobel prize in Chemistry in 1932 for this work. After that, many studies on adsorption have been carried out and confirmed that adsorption processes would play a fundamental part in the reaction mechanism of heterogeneous reactions.

5.1　General aspects of the adsorption process

This introductory section aims to define some basic and general terms that will be used later. Adsorption is a phenomenon in which solutes and gas molecules are removed from the liquid or gas phase to the solid or liquid surface owing to the surface forces. The origin of the surface force was explained in Section 1.4.1. The relative energies of the surface atoms are higher than those in the bulk atoms. Therefore, atoms at the surface seek other atoms for stabilization.

This process creates the accumulation of many molecular species at the surface compared to the bulk. The substance that adsorbs another substance is called adsorbent, and the adsorbed substance is called adsorbate. The adsorption process is discussed through graphs known as **adsorption isotherms**, which shows the relationship between the amounts of adsorbate on the surface of the adsorbent and pressure at a constant temperature. Desorption is a phenomenon in which a substance is released from a surface. The process is the opposite of adsorption. The adsorption phenomenon is schematically shown in Fig. 5-1.

It is necessary to evaluate the amount of the adsorbed species at the surface or interface to understand the adsorbed species' roles. Due to the recent development of surface analytical

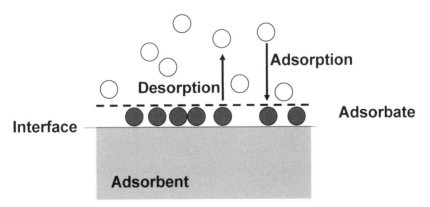

Fig. 5-1　Schematic view of adsorption and desorption process.

instruments such as ESCA, EBMA, AES, LEED, and AFM, we can now get very detailed information about the solid surface and the adsorption structure of precise adsorption behaviors can be obtained at the atomic level.

How does adsorption occur?

Adsorption occurs because of the attraction forces between an approaching molecule and a surface. These forces can be attributed to two types of interactions. The first is the bonds which are established between the adsorbate and the adsorbent. These are known as Van der Waals or dispersion forces, and they are introduced instantaneously by induced dipole moments between an approaching molecule and a surface. The bond energy is low, typically less than 20 kJ mole^{-1}. This type of adsorption is called physical adsorption or "physisorption." It generally occurs at relatively low temperature

The second type of adsorption is due to the chemical bonds between the adsorbate and adsorbent, or the electron exchanges which occur. The energies here are generally much higher and of a similar order of magnitude in ordinary chemical reactions, that is, 300-500 kJ mole^{-1}. This adsorption is called chemical adsorption or "chemisorption."

Heat of adsorption

All of the adsorption processes are exothermic reactions. That is, adsorption always produces heat. Consider a gas molecule adsorption. The gas molecule is removed from the gas phase and moves to the interface in the adsorption process. Before the adsorption, the gas molecule is moving around in 3-dimensional space. After adsorption, the adsorbed molecule is constrained in the 2-dimensional surface. From the thermodynamical viewpoint of this change, the entropy of a molecule decreases. Namely, ΔS is always negative ($\Delta S < 0$) in the adsorption process. The adsorption occurrence means that the free energy ΔG must be decreased by adsorption, or ΔG of the adsorptions must be negative.

$$\Delta G = \Delta H - T\Delta S < 0 \qquad ----- (5\text{-}1)$$

ΔH is enthalpy change, and T is the absolute temperature. Since ΔS is always negative or $\Delta S < 0$. Then,

$$\Delta H < T\Delta S < 0 \qquad ----- (5\text{-}2)$$

The negative value of ΔH means the adsorption is an exothermic process. That is, **adsorption always generates heat**. The heat generation based on the adsorption is called "heat of adsorption." In general, physical adsorption generates heat of about 10 kJ mol^{-1}, but chemical adsorption reaches 50 to 500 kJ mol^{-1}.

When the reaction gas is introduced into the sample surface at a high temperature, it is unnecessary to heat the introducing gas because when the reaction gas molecules are adsorbed on the sample surface, the heat is generated due to the heat of adsorption. The raised temperature due to the adsorption quickly dissipates to the sample and returns to the preset temperature. However, in the external heating system, the surface temperature may decrease due to increased flow rate, so it is necessary to control the surface temperature not to drop.

5.2 How to measure the amount of adsorption

When adsorption occurs at the gas-solid interface, the weight of a solid sample increases, and the pressure of the gas decreases in a constant volume vessel. Thus, the amount adsorbed can be measured in at least two ways: measuring the change in weight of the solid sample with a balance or measuring the change in pressure of the gas in an accurately known volume. These techniques are termed gravimetric and volumetric determination, respectively.

Gravimetric and volumetric methods can be effectively used to study the behavior of adsorption on solid surfaces. In general, however, these cannot be applied for the measurements of solute adsorption at liquid or solution surfaces since liquids have relatively large solubility of molecules. It is hard to separate the amount of adsorption and dissolution from gravimetric and volumetric measurements. For the evaluation of adsorption at liquid surfaces, other methods must be applied. (See Section 5-5)

5.3 Langmuir adsorption isotherm

The adsorption of a homogeneous surface can be described by the Langmuir adsorption isotherm given by Langmuir in 1916. In the model, all adsorption sites on the surface are the same, and no interactions between one adsorbed molecule and another Thus, it is sometimes called ideal Langmuir adsorption. Only a monolayer can be adsorbed and multi-layer adsorption can be neglected.

At first, consider adsorption at the gas-solid surface, assuming the homogeneous solid surface and the gas phase consists of one component. If the surface is initially bare of adsorbed species, the adsorption begins when the surface is exposed to gas at a given temperature T and pressure P. The adsorption rate is proportional to the flux of molecules to the surface (which in turn is proportional to the pressure) and to the fraction of unoccupied sites $(1 - \theta)$. θ is the fraction of the surface covered by adsorbed molecules. Thus,

$$\text{Rate of adsorption} = k_a P(1 - \theta) \qquad ----- (5\text{-}3)$$

where k_a is a proportional constant. The rate of desorption is proportional to the fraction θ.

$$\text{Rate of desorption} = k_d \theta \qquad ----- (5\text{-}4)$$

where k_d is a proportional constant. *At equilibrium*, the rates of adsorption and desorption are equal. Setting the two rates equal to each other and solving for θ, we obtain the following equations,

$$\frac{\theta}{1 - \theta} = KP \qquad ----- (5\text{-}5)$$

$$\theta = \frac{KP}{1 + KP} \qquad ----- (5\text{-}6)$$

78

where $K = k_a/k_d$ is called adsorption constant. This equation (5-5) is known as the Langmuir adsorption isotherm or an ideal Langmuir adsorption isotherm since it neglects the interaction among adsorbed species. The adsorption behavior is demonstrated in the plot of θ versus P shown in Fig. 5-2.

When P is sufficiently low ($KP \ll 1$), the fraction of occupied sites increases linearly with P, or $\theta \approx KP$, while it approaches unity, $\theta \approx 1$, when P is large enough for $KP \gg 1$. These are shown in Fig. 5.2 as broken lines.

Adsorption decreases with increased temperature since the bonding between adsorbed species and the surface atoms becomes weak due to increased thermal vibration. Thus, adsorption can be very small at high temperatures, such as more than 1000 K. In these cases, it is reasonably assumed that $\theta \approx KP$. This means that the surface coverage is proportional to the pressure. The surface coverage and pressure are proportional to the surface concentration and bulk concentration, respectively. **Thus, the surface concentration is proportional to the bulk concentration at a relatively high temperature**.

In the rate law expression for heterogeneous reactions, we need to use exact values for the surface concentration. However, for the reason mentioned above, it is possible to use bulk concentrations instead of surface concentrations at high-temperature heterogeneous reactions.

Initially, the Langmuir model was derived to describe adsorption isotherm for gases on solids, but it has since been realized that the model is also applicable to solute adsorption in many solution systems. For solute adsorption, concentrations can be used instead of pressure.

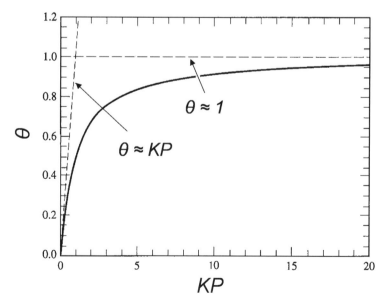

Fig. 5-2 Langmuir adsorption isotherm.

In this case, isotherm (5-6) is expressed by concentrations,

$$\theta = \frac{K[A]}{1 + K[A]}$$ ----- (5-6)'

where [A] is the concentration of the solute A. For example, it is well known that the sulfur (S) adsorption on the molten Fe-S surface is described by the Langmuir adsorption isotherm with the dissolved sulfur concentration. The Langmuir model has been used as the starting point for the derivation of isotherm for more complex systems such as competitive adsorption, dissociative adsorption and multi-stage adsorption.

5.4 Extension of the Langmuir adsorption isotherm

5.4.1 Competitive adsorption

It is known that two or more species can occupy the sites at a surface or interface simultaneously. For example, in molten Fe-O-S alloy, it is known that dissolved O and S adsorb at the surface simultaneously.

Suppose the melt containing solutes A and B. Solute A has fractional coverage θ_A and second solute B has coverage θ_B. Then the rates for adsorption of solutes A and B are $k_a^A (1 - \theta_A - \theta_B)[A]$ and $k_a^B (1 - \theta_A - \theta_B)[B]$, respectively, since in each case, the rate of adsorption is proportional to the fraction of unoccupied free sites $(1 - \theta_A - \theta_B)$. k_a^A and k_a^B are the adsorption rate constants of A and B, respectively. Similarly, the rate of desorption in each case is proportional to the number of sites occupied by each solute. For A and B, these rates are $k_d^A \theta_A$ and $k_d^B \theta_B$, respectively. K_d^A and k_d^B are desorption rate constants of A and B, respectively.

At equilibrium, the rates of adsorption and desorption for each species must be equal. Denoting $K_A = k_a^A / k_d^A$ and $K_B = k_a^B / k_d^B$, we find

$$\frac{\theta_A}{1 - \theta_A - \theta_B} = K_A[A]$$ ----- (5-7)

$$\frac{\theta_B}{1 - \theta_A - \theta_B} = K_B[B]$$ ----- (5-8)

After a little algebra, we find that

$$\theta_A = \frac{K_A[A]}{1 + K_A[A] + K_B[B]}$$ ----- (5-9)

and that

$$\theta_B = \frac{K_B[B]}{1 + K_A[A] + K_B[B]}$$ ----- (5-10)

In the extreme limit when $K_A[A] \gg K_B[B]$, B is almost totally excluded from the surface, while the surface coverage of A approaches the value it would have in the absence of B.

5.4.2 Dissociative adsorption

Dissociative adsorption is critically important for heterogeneous reactions. Consider the following reaction of $A_2 + B_2 = 2AB$. Even if this reaction does not occur in the homogeneous reaction system, it can sometimes proceed in a heterogeneous reaction system. This is simply due to dissociative adsorptions. For example, while gaseous N_2 is generally an inert or unreactive molecule, it becomes reactive when converted into two reactive radicals, such as $2N$. The dissociative adsorption requires two available sites. The desorption occurs as a bimolecular step involving the combination of two surface atoms.

The general dissociative adsorption of a homo-nuclear diatomic molecule, A_2, can be represented by the following chemical equation:

$$A_2(g) = 2A(ads)$$

The rate of adsorption is proportional to the flux of molecules to the surface (which in turn is proportional to the pressure P_{A_2}) and to the fraction of unoccupied sites $(1 - \theta_A)^2$. θ_A is the fraction of the surface covered by adsorbed molecules. Thus,

$$\text{Rate of adsorption} = kaP_{A_2} (1 - \theta_A)^2 \qquad \text{----- (5-11)}$$

The rate of desorption is proportional to the fraction θ_A^2.

$$\text{Rate of desorption} = k_d \theta_A^2 \qquad \text{----- (5-12)}$$

Equating the two rates gives the following expression for the isotherm for dissociative adsorption:

$$\frac{\theta_A^2}{(1 - \theta_A)^2} = KP_{A_2} \qquad \text{----- (5-13)}$$

or

$$\theta_A = \frac{\sqrt{KP_{A_2}}}{1 + \sqrt{KP_{A_2}}} \qquad \text{-----(5-14)}$$

where $K = k_a/k_d$. We have an isotherm that is formally analogous to that of Langmuir, but in which P is replaced by $(P)^{1/2}$

5.5 Solute adsorption

5.5.1 Gibbs adsorption equation

In the experimental section, a gravimetric and volumetric method to measure the adsorbed amounts were introduced. However, these are hard to apply for the measurements of solute adsorption at solution surfaces since the liquid has a relatively large solubility of molecules. For the evaluation of adsorption at the solution surface, another method must be applied. One of the most effective methods to study solute adsorption behaviors at a liquid surface is to measure the surface tension.

The Gibbs adsorption equation can evaluate the amount of adsorbed species at the surface from the change of the surface tension. The Gibbs adsorption equation for a multicomponent solution is expressed by

$$d\gamma = - \sum_i^n \Gamma_i d\mu_i \qquad \text{----- (5-15)}$$

where γ is surface tension, μ_i is the chemical potential of component i, and Γ_i is surface excess of component *i* per unit surface area. Surface excess is a concept proposed by J. W. Gibbs to handle interfaces and surfaces thermodynamically. Qualitatively speaking, it can be simply regarded as the amount of solute adsorbed on the surface for the Fe-O and Fe-S systems. Details of Gibbs adsorption equation and the surface excess can be found elsewhere [1].

For a binary system at constant temperature eq. (5-15) reduces to

$$d\gamma = -\Gamma_1 \, d\mu_1 - \Gamma_2 \, d\mu_2 \qquad \text{----- (5-16)}$$

Components *1* and *2* correspond to the solvent and solute, respectively. Since Γ_1 and Γ_2 are defined relative to an arbitrarily chosen dividing surface plane, it is possible to position the surface so that $\Gamma_1 = 0$. Namely, no excess amount of solvent at the surface. Then,

$$\Gamma_2 = -(\partial_\gamma/\partial\mu_2)$$

or $\qquad \Gamma_2 = -(d\gamma/dRT \ln a_2) = -(a/RT)(d\gamma/da_2) \qquad \text{----- (5-17)}$

where R is the gas constant, T is the absolute temperature and a_2 is the activity of the solute. It is explained based on the binary Fe(l)-O system to simplify the explanation. Then, eq. (5-17) is,

$$\frac{d\sigma}{d \ln a_O} = RT \, \Gamma_O \qquad \text{----- (5-18)}$$

σ is the surface tension of liquid Fe, a_O is the oxygen activity in Fe(l), Γ_O is the amount of adsorbed oxygen at the liquid Fe surface.

Eq. (5-18) means that the amount of adsorbed oxygen at a particular oxygen concentration can be obtained from the differentiation of the surface tension by the chemical potential at a particular oxygen concentration.

Typical experimental results of surface tension variation of Fe-O system with oxygen con-

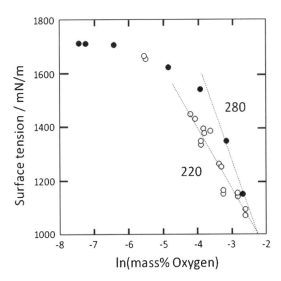

Fig. 5-3 Effect of oxygen content on the surface tension of Fe-O system. Closed circles by Kozakevitch and Urbain [2] and open circles by Halden and Kingery [3].

tent by Kozakevitch and Urbain [2] and Halden and Kingery [3] are shown in Fig. 5-3. The measured data are somewhat scattered, although these measurements were supposed to be one of the most precise measurements. Therefore, it is practically impossible to calculate the accurate, reliable differential values of $d\sigma/d\ln a_O$ from these surface measurements. The numeric values in the figure correspond to the $d\sigma/d\ln a_O$ near the oxygen-saturated region.

5.5.2 Belton equation

It is generally difficult to calculate reliable differential values for surface tension measurements. G. Belton proposed the following equation to conquer this problem: the equation combines the Gibbs adsorption equation and Langmuir adsorption isotherm equation [4]. These two equations are presented for ease of explanation, with the molten Fe-O system as an example.

$$\frac{d\sigma}{d\ln a_O} = RT\,\Gamma_O \qquad\qquad \text{----- (5-18)}$$

$$\frac{\theta_O}{1 - \theta_O} = K_O\,a_O \qquad\qquad \text{----- (5-19)}$$

a_O is the activity of dissolved oxygen in Fe, θ_O is fractional coverage of adsorbed O at the surface and expressed by $\theta_O = \Gamma_O/\Gamma_O^0$, and will be a unit at the saturation. Γ_O^0 is the saturated oxygen content in liquid Fe surface. K_O is the adsorption equilibrium constant for oxygen. It is found that the Langmuir adsorption isotherm equation can describe the behavior of the adsorbed oxygen at the liquid surface. Combined with (5-18) and (5-19)

$$\sigma^P - \sigma = -RT\,\Gamma_O^0 \ln\left(1 + K_O\,a_O\right) \qquad\qquad \text{----- (5-20)}$$

σ^P is the surface tension for pure liquid Fe. Unlike the eq. (5-18), equation (5-20) expresses the variation of the surface tension value itself but not differential values of the surface tension. In other words, the Belton equation is an integrated form, but the Gibbs equation is a differentiated form. Thus, eq. (5-20) is much less sensible for use with experimental fluctuations.

Eq. (5-20) contains two parameters of K_O and Γ_O^0. As can be seen from Fig. 5-3, the gradient, $d\sigma/d\ln a_O$, hardly changes in the high oxygen concentration region and regarded to be constant. This value can be considered as Γ_O^0 (the saturated oxygen content). Then, by fitting the surface tension data with eq. (5-20) by varying K_O with constant Γ_O^0, the value of K_O can be evaluated. By using the surface depression results of Kozakevitch and Urbain [2] at 1823 K with taking the saturation coverage derived by Kozakevitch (1968) [5], the best fitting yields:

$$\sigma = 1788 - 281\ln[1 + 142(\text{mass\%O})] \quad (\text{mN/m}) \quad \text{----- (5-21)}$$

The fitting result (solid line) and experimental results (open circles [1]) are shown in Fig. 5-4. There is in close accord with the results. Thus, the oxygen adsorption equilibrium constant K_O is 110 (1 mass % standard for a_O). In this way, it is practically possible to calculate the adsorption constants, K_i of adsorbed species i from the surface tension dependency with solute concentrations by applying the Belton equation. There is, however, still a problem with the

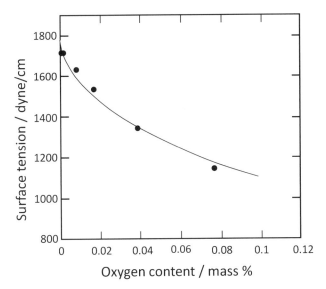

Fig. 5-4 The depression of the surface tension of iron by oxygen and comparison of the experimental results by Kozakevitch and Urbain [2] with the isotherm eq. (17) at 1823 K.

reliability of the experimental measurements since K_i is strongly dependent on the precise measurement of the surface tension.

From eq. (5-18) or (5-20), RT $\Gamma_O^0 = 281$. Then, $\Gamma_O^0 = 1.82 \times 10^{-9}$ (mole/cm²) = 1.08×10^{15} (atoms/cm²) at 1823 K. Then an adsorbed oxygen area can be $1/1.152 \times 10^{15} = 9.15 \times 10^{-16}$ cm². Then, the calculated adsorbed O radius is about 17.0 nm. This value is reasonably close to the ionic radius of O^{2-} ions of about 14 nm. It means that the structure of adsorbed oxygens on the molten Fe surface may be close to the oxygen ions.

5.6 Evaluation of adsorption behaviors from reaction kinetics

Adsorption behaviors in solutions can be quantitatively measured from the dependency of surface tensions variation with dissolved element concentrations. Another effective method to evaluate the adsorption behaviors is to use the reaction rates. It is well known that Chalcogen elements (VI group in the periodic table), such as O and S, strongly adsorb at the surface, leading to a reduction in the available reaction sites. They are called "surface-active elements".

Let's think about oxygen adsorption for the molten Fe-O system. A portion of the reaction surface is occupied by adsorbed O (O(ad)). When k_0 expresses the reaction rate constant at the bare surface, and that with available reaction site with θ_O is defined by k_a. In this case, the available reaction area is $(1 - \theta_O)$, and the apparent reaction rate is reduced by $(1 - \theta_O)$. If the reaction condition does not change except the reaction area, then the following relation is established.

$$k_a = k_0 (1 - \theta_O)$$
----- (5-22)

If the adsorbed oxygen behavior is reasonably expressed by ideal Langmuir isotherm, by combined eq. (5-19) with (5-22),

$$k_a = \frac{k_0}{1 + K_O a_O}$$
----- (5-23)

where a_O is the activity of oxygen. Eq. (5-23) can be modified into

$$\frac{1}{k_a} = \frac{1}{k_0} + \frac{K_O a_O}{k_0}$$
----- (5-24)

Suppose the plot of $1/k_a$ versus a_O will show a linear relationship, as shown in Fig. 5-5. K_O can be evaluated from the intersection (k_0) and the slope (K_O/k_0). For the evaluation of the adsorption constant K_O by applying the eq. (5-24), we have to know the rates with various amounts of O(ad).

5.7 Role of adsorbed oxygen to determine pO_2

As already mentioned, oxygen is a strong surface-active element, and adsorbed oxygen blocks the reacting sites to reduce the apparent reaction rates. The blocking oxygen plays a vital role in determining the oxygen potential in the heterogeneous reacting system.

To fix the oxygen activity or oxygen potential in molten Fe, Fe is equilibrated with a CO-CO_2 gas mixture with a particular CO_2/CO ratio. Under the equilibrated condition, the following two reactions occur.

$$CO_2(g) = CO(g) + O_2/2(g)$$
----- (5-25)

$$O_2/2(g) = \underline{O}$$
----- (5-26)

where $\underline{O}$ is the dissolved oxygen in molten Fe. Since the reaction (5-25) is under equilibria, the oxygen potential can be expressed by

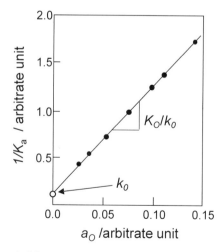

Fig. 5-5 Reciprocal of the apparent rate constants 1(/ka) at 1823 K as a function of a_O.

$$\sqrt{pO_2} = K_{25} \frac{pCO_2}{pCO} \qquad\qquad ----- (5\text{-}27)$$

where K_{25} is the equilibrium constant of the reaction (5-25). Namely, once the pCO_2/pCO ratio is fixed, the pO_2 can be determined.

The reaction (5-26) is also under the equilibrium condition. The oxygen activity in molten Fe, aO, can be expressed by the following equation,

$$\sqrt{pO_2} = K_{26} a_O \qquad\qquad ----- (5\text{-}28)$$

where K_{26} is the equilibrium constant of the reaction (5-26). Combined the equation (5-27) and (5-28),

$$a_O = \frac{K_{25}}{K_{26}} \frac{pCO_2}{pCO} \qquad\qquad -------(5\text{-}29)$$

From equation (5-29), a_O can be decided by fixing the pCO_2/pCO ratio. Using equation (5-29), the a_O at 1873 K with the constant CO_2/CO ratio of 0.1 is calculated to about 10^{-14} atm under the total pressure of 1 atm. Also, the dissolved oxygen content is 50 ppm. From the thermodynamic viewpoint, the a_O in liquid Fe is simply achieved by equilibrating liquid Fe with a CO_2-CO gas mixture of a particular pCO_2/pCO ratio.

Suppose that liquid iron is kept under high vacuum conditions with a small amount of O_2 of its partial pressure of 10^{-10} atm at 1723 K using a vacuum pump, then the dissolved oxygen content in Fe will be 50 ppm (Fig. 5.6 (a)). Practically, however, it is very hard to attain this high vacuum even by using the most highly developed and high performance pump of our time. In contrast, $\underline{O}$ content of 50 ppm easily attained without a high vacuum condition but by introducing $CO\text{-}10\%CO_2$ mixture of the total pressure of 1atm to liquid Fe. (Fig. 5-6 (b)).The $\underline{O}$ content is calculated using the following thermodynamic data,

$$O_2/2 = \underline{O} \qquad\qquad \varDelta G^0 = \qquad kJ/mol$$
$$FeO(l) + CO = Fe(l) + CO_2 \quad \varDelta G^0 = -24.87 + 0.031T \qquad kJ/mol$$

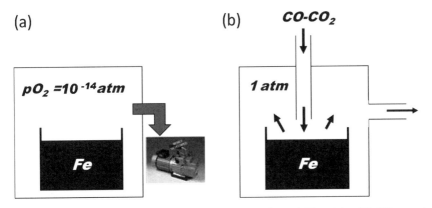

Fig. 5-6   Establishment of low oxygen pressure by (a) evacuation and (b) applying CO-CO$_2$ gas mixture.

The pO_2 is established through the reactions of (5-25) and (5-26). However, thermodynamics does not tell the detailed reaction mechanism of the pO_2 fixing process. The process can be understood by taking the existence of adsorbed oxygen into account.

Now let us think about this process from the kinetics viewpoint. In the reaction between liquid Fe and CO-CO_2 gas mixture, there are adsorbed oxygen atoms, O(ad), at the surface. Under the equilibrium condition, O(ad) is equilibrated with the dissolved oxygen in liquid Fe.

$$O(ad) = \underline{O} \qquad \text{----- (5-30)}$$

The O(ad) is supplied by the dissociation reaction of CO_2 (oxidation process) and removed from the surface by the reaction between CO and O(ad) (reduction reaction). The following equations can express these reactions,

$$CO_2(g) \rightarrow O(ad) + CO(g) \qquad \text{----- (5-31)}$$
$$CO(g) + O(ad) \rightarrow CO_2(g) \qquad \text{----- (5-32)}$$

The reactions of (5-31) and (5-32) can be separated into several elementary reactions. Since the essence of the explanation does not change by neglecting these elementary steps, they can be ignored.

Under equilibrium, the amount of O(ad) does not change. Thus, the supply rate and removal rate of O(ad) must be equal. It is well known that O(ad) is strongly absorbed on the reaction sites of liquid Fe surface and block the CO_2 gas absorption on the reaction sites. Therefore, the amount of CO_2 absorption decreases by the increase of O(ad) atoms. Consequently, the CO2 dissociation reaction at the surface decreases, or the O(ad) supply rate decreases. Reversely, the removal rate of O(ad) by CO increases with O(ad). These situations are schematically shown in Fig. 5-7. Line A and B indicate the supply rate of O(ad) and removal rate of O(ad) as a function of O(ad), respectively. Each reaction rate has the same value at the intersection (E) of the two lines. At this point, the removal and supply rates of O(ad) are equal. This situation corresponds to the equilibrium state.

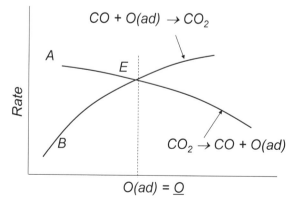

Fig. 5-7 The equilibrium oxygen partial pressure establishing mechanism by CO_2-CO gas mixture.

A small ratio of CO_2/CO must be used to establish a very low pO_2 using a CO-CO_2 gas mixture. With an increase in pCO, however, carbon deposition begins. Therefore, for a low pO_2 setting, H_2O-H_2 gas is often used instead of CO-CO_2. The pO_2 determination process by H_2O-H_2 is essentially the same as that for CO-CO_2. That is, the pO_2 is simply determined by the balance between the supply rate of $O(ad)$ by H_2O and the removal rate of $O(ad)$ by H_2.

pO_2 is also determined by using CO_2-H_2 or CO-H_2O gas mixture. In these gas mixtures, the pO_2 is also determined by the balance between the $O(ad)$ supply rate and removal rate under the equilibrium condition like CO_2-CO or H_2O-H_2. Unlike the gas mixtures of CO_2-CO or H_2O-H_2, however, in the case of CO_2-H_2 or CO-H_2O gas mixture, after reaching to equilibrium condition, their equilibrium gas compositions will consist of 4 gas species of CO_2, CO, H_2, and H_2O. Under the equilibrium condition, the following two reactions should also be under equilibrium:

$$CO_2\ (g) = CO(g) + O_2/2(g) \qquad \text{----- (5-25)}$$
$$H_2O\ (g) = H_2(g) + O_2/2(g) \qquad \text{----- (5-33)}$$

The oxygen pressure pO_2 determined by the reaction (5-25) and (5-33) must be equal. This equilibrium situation is illustrated in Fig. 5-8.

It is noted that the introduced gas mixture must be fully converted to the equilibrium compositions when CO_2-H_2 or CO-H_2O gas mixture is used to fix pO_2. To achieve this, a catalyst, such as a bundle of Pt wire, is often inserted in the gas delivery tube with a reasonably slow gas flow rate to assure the full conversion of the introduced gas.

5.8 Steady-state pO_2 with H_2-CO gas mixture

CO-H_2O or CO_2-H_2 gas mixtures can be used to establish the equilibrium pO_2. However, in these cases, the introducing gas mixture must be fully converted into the equilibrium gas composition to establish the equilibrium pO_2. What happens, however, if the full conversion

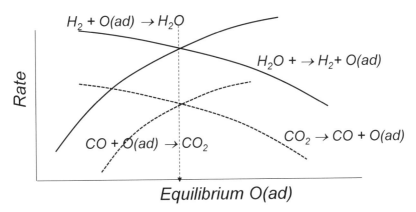

Fig. 5-8 The equilibrium oxygen partial pressure establishing mechanism by CO_2-H_2 or CO-H_2O gas mixture.

is retarded by introducing a high flow rate so that CO_2 or H_2 formation (in the case of $CO-H_2O$) becomes negligibly small? This situation corresponds to point S in Fig. 5-9.

The obtained pO_2 is not the equilibrium one but "the steady-state" one. Under this steady-state condition, the Oad supply rate and removal rate are expressed by,

$$\text{Supply rate} = k_1 pH_2O (1 - \theta_O) \qquad ----- (5-34)$$
$$\text{Removal rate} = k_2 pCO\theta_O \qquad ----- (5-35)$$

Assuming Langmuir adsorption isotherm,

$$\theta_O/1 - \theta_O = K_O a_O \qquad ----- (5-19)$$

where K_O is adsorption constant, and a_O is the steady-state oxygen activity and defined by

$$a_O = (pO_2)^{1/2}. \qquad ----- (5-36)$$

Under the steady-state, the rates of (5-34) and (5-35) are equal. Combination of (5-19), (5-34), (5-35), and (5-36) yields,

$$k_1/k_2 = a_O K_O/(pH_2O/pCO) \qquad ----- (5-37)$$

The value of K_O and CO_2 dissociation rate coefficient k_1 are known. Therefore, once the steady-state pO_2 is measured, k_2 can be calculated by applying Eq. (5-36).

Sasaki and Belton measured the steady-state oxygen potential in Fe by introducing an H_2O-CO gas mixture [6]. The measured steady-state oxygen potential with various H_2O-CO gas mixtures and the calculated values for complete water-gas equilibrium for these gas mixtures at 1973 K are shown in Fig. 5-10.

The values for k_2/k_1 are calculated to be 2.6 at 1973 K (1700°C) from the slope shown in Fig. 5-10. This result means that the decomposition rate coefficient of H_2O can be evaluated from the decomposition reaction rate of CO_2 on the surface of molten iron by measuring the steady oxygen potential with H_2O-CO mixed gas. This result demonstrates the transferability of **elementary reaction** rate constants from one mechanism to another. In this sense, chemical kinetics is a predicting science. In other words, elementary reactions in kinetics play the same role comparable to the pure substance in the tabulation of thermodynamic data.

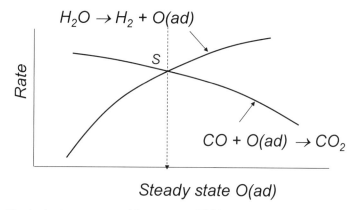

Fig. 5-9 The steady-state oxygen partial pressure establishing a mechanism by CO-H₂O gas mixture.

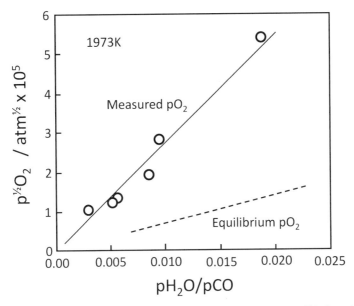

Fig. 5-10 Comparison of measured steady-state oxygen activity presented by $p^{1/2}O_2$ in molten Fe as a function of pH_2O/pCO at 1973 K, with the calculated values for complete water-gas equilibrium (indicated by broken lines) [6].

Further reading

A. W. Adamson: Physical chemistry of Surfaces 3rd ed. , John Wiley & Sons, New York, 1976.

D. K. Chattoraj and K. S. Birdi: Adsorption and the Gibbs Surface Excess, Plenum Press New York, 1984.

R. Aveyard and D. A. Haydon: An introduction to the principles of surface chemistry, Cambridge University Press, 1973.

M. Prutton.: Surface Physics 2nd ed., Clarendon press, Oxford 1983.

References

[1] A. W. Adamson: Physical Chemistry of Surface 3rd ed., John Wiley & Sons, New York, 1976, 68-81.

[2] P. Kozakevitch and Urbain: Mem. Sci. Rev. Metall.,58 (1961), 517-534.

[3] F. A. Halden and W. D. Kingery: J. Phys. Chem. 59 (1955), 557-559.

[4] G. R. Belton: Met. Trans B., 7B (1976), 35-42.

[5] P. Kozakevitch: Surface phenomena of metals, Monograph No. 28 of the society of chemical industry London, 1968, 223-45

[6] Y. Sasaki and G. R. Belton: Metallurgical and Materials Transactions B, (1998), pp.829-835.

Applications

Chapter 6
Decarburization of Fe-C by CO_2

Chapter 6   Decarburization of Fe-C by CO$_2$

6.1   CO$_2$ dissociation reaction

The kinetics of the decarburization of liquid Fe-C by oxidizing gases has been extensively studied due to its practical importance in steelmaking. Most of the studies of the decarburization reaction of Fe-C have been carried out using CO$_2$ or CO-CO$_2$ gas mixture because of the extremely fast decarburization reaction rate of molten Fe-C with O$_2$ gas. The overall decarburization reaction of molten Fe-C by CO$_2$ is given as

$$CO_2(g) + \underline{C} \rightarrow 2CO(g) \qquad\qquad \text{----- (6-1)}$$

This reaction (6-1) consists of several steps,

$$CO_2(g) \rightarrow CO_2(ad) \qquad\qquad \text{----- (6-2)}$$

$$CO_2(ad) \rightarrow O(ad) + CO(ad) \qquad\qquad \text{----- (6-3)}$$

$$\underline{C} \rightarrow C(ad) \qquad\qquad \text{----- (6-4)}$$

$$O(ad) \rightarrow \underline{O} \qquad\qquad \text{----- (6-5)}$$

$$C(ad) + O(ad) \rightarrow CO(ad) \qquad\qquad \text{----- (6-6)}$$

$$CO(ad) \rightarrow CO(g) \qquad\qquad \text{----- (6-7)}$$

In the decarburization reaction of the Fe-C melt with an oxidizing gas such as CO$_2$, the carbon content in the melt gradually decreases along with the reaction. Because the reaction (6-6) is quite fast (the details will be discussed in Chapter 7), the reaction (6-6) is assumed to be under the equilibrium condition. Therefore, the C(ad) decrease results in an increase in O(ad) at the melt surface during the decarburization reaction. The increase in O(ad) decreases the decarburization reaction rate since O(ad) blocks the reaction sites. Experiments without any change in the O(ad) content can be carried out by applying carbon-saturated conditions. Because the amount of O(ad) is constant and very small under the carbon saturated condition, the surface is regarded as a bare surface on which no O(ad) blocks the adsorption sites.

Sain and Belton [1] carried out decarburization experiments of Fe-C melts with carbon-saturated conditions. They designed the system to avoid mass transfer limitations by using an inductively stirred melt and an impinging gas jet with very high gas flow rates of up to 40 l/min to the liquid Fe surface, forcing the system into a regime of interfacial reaction control. For the carbon-saturated condition, a graphite disc was cemented to the bottom of the crucible. A high purity Fe sample (with a sulfur content of less than 0.0005 mass%) was used to minimize or avoid the reaction site blockage by adsorbed S atoms. The reduction rate was evaluated from the weight loss. The experimental arrangement of reacting part is shown in Fig. 6-1.

Their results of the decarburization rates of high purity Fe with carbon saturation as a function pCO$_2$ for several temperatures are presented in Fig. 6-2.

From a linear dependency on the CO$_2$ partial pressure, as shown in Fig. 6-2, the decarburization rate is presented by

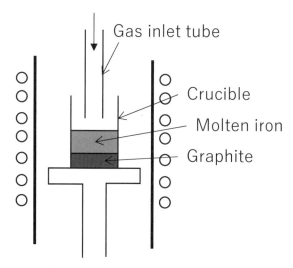

Fig. 6-1   The experimental arrangement for decarburization of liquid iron by CO_2 [1].

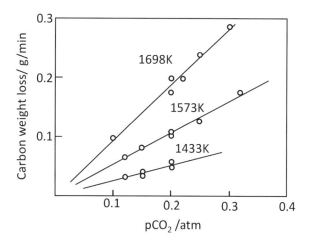

Fig. 6-2   Dependence of the decarburization rates with carbon saturated condition in CO_2-CO mixture at several temperatures [1].

$$d[C]/dt = -k_f pCO_2 \qquad\qquad \text{----- (6-8)}$$

where k_f is the forward rate constant (mol/(cm^2 s atm)) at the bare surface, [C] is the carbon mol% concentration in the liquid Fe. From the temperature dependency, the rate constant k_f as a function of temperature is expressed by eq. (6-9)

$$\log k_f = -5080/T - 0.21 \qquad\qquad \text{----- (6-9)}$$

Since the decarburization rate was solely determined by pCO_2 as shown in Fig. 6-2, pCO does not affect the rate. The independence of the rate on pCO indicates that the chemisorption of CO plays no major role. That is, the rate-controlling step of the decarburization reaction

(6-1) is either (I) the rate of chemisorption of CO_2

$$CO_2(g) \rightarrow CO_2(ad) \qquad\qquad ----- (6\text{-}2)$$

or (II) the rate of the dissociation of adsorbed CO_2 on the Fe surface

$$CO_2(ad) \rightarrow O(ad) + CO(ad) \qquad ----- (6\text{-}3)$$

Since the rate equation of (6-2) or (6-3) has the same rate expression of eq. (6-8), it is not possible to tell which reaction of (6-2) or (6-3) is the rate-controlling step. Thus, the rate-controlling step of the decarburization reaction is (6-2) or (6-3).

When expressing the reaction rate equation for a reaction, the rate equation changes depending on what kind of raw material or product is focused on, and the unit of the reaction rate constant also changes accordingly. Concerning the decrease of carbon weight loss, the rate equation of the decarburization reaction of (6-1) can be expressed by

$$d[C]/dt = -k_f pCO_2 \qquad\qquad ----- (6\text{-}8)$$

When focussed on the change of the CO_2 concentration (mol/cm³), the rate equation will be

$$d[pCO_2]/dt = -k_f pCO_2 \qquad ----- (6\text{-}8)'$$

The unit of k_f is mol/(cm² s atm) for (6-8) and 1/s for (6-8)'. When comparing results with other researchers, one must pay attention to the unit of reaction rate constant.

6.2   The effect of O adsorption

When the Fe-C melt is not carbon saturated, there will be some dissolved oxygen in the melt. Due to the reaction (6-5)',

$$\underline{O} \rightarrow O(ad) \qquad\qquad ----- (6\text{-}5)'$$

O(ad) exists on the surface in the Fe-C melts when the carbon saturated condition is not met. O(ad) strongly influences the decarburization reaction of Fe-C melts (not saturated with C) since O(ad) blocks the reaction sites. For the evaluation of the effect of O(ad) on the decarburization of Fe-C melt, the isotope exchange method is an attractive approach since the reactions can be carried out under the equilibrium condition.

For the reaction of liquid Fe with the CO_2-CO gas mixture, the oxygen $\underline{O}$ and carbon concentration $\underline{C}$ in liquid Fe is controlled by the following reactions.

$$CO_2(g) \rightarrow CO(g) + \underline{O} \qquad ----- (6\text{-}10)$$
$$CO(g) \rightarrow \underline{C} + \underline{O} \qquad\qquad ----- (6\text{-}11)$$

Combination of (6-10) and (6-11) gives

$$CO_2(ad) + \underline{C} \rightarrow 2CO(ad) \qquad ----- (6\text{-}12)$$

Under the equilibrium condition of Fe-C melt with CO2-CO gas mixtures, the oxygen $\underline{O}$ concentration is fixed by (6-10) or pCO_2/pCO and the carbon concentration $\underline{C}$ is fixed by (6-12) or $(pCO)^2/pCO_2$.

The rate-determining step in the following isotope exchange reaction,

$$^{14}CO_2 + {}^{12}CO \rightarrow {}^{14}CO + {}^{12}CO_2 \qquad \text{----- (4-6)}$$

is found to be

$$^{14}CO_2(g) \rightarrow {}^{14}CO(g) + O(ad) \qquad \text{----- (4-7)}$$

The dissociation reaction behavior of $^{14}CO_2$ can be practically identical to $^{12}CO_2$. Thus, by measuring the production rate of ^{14}CO, the rate of untagged reaction (6-13) can be evaluated.

$$CO_2 \rightarrow CO + O(ad) \qquad \text{----- (6-13)}$$

Cramb and Belton [2] carried out the isotope exchange reaction of (4-6) on liquid Fe-C surface with varying pCO_2/pCO ratios at various temperatures. Unlike the saturated condition, in the isotope exchange reaction, the surface coverage can be changed by varying pCO_2/pCO ratios.

As mentioned in section 4.3.1, the following equation can calculate the reaction rate constant k of the reaction (4-7).

$$k = \frac{V_t}{ART} \frac{1}{1+B} \ln \left[\frac{1}{1 - P^{14}CO/(P^{14}CO)_{eq}} \right] \qquad \text{----- (4-11)}$$

where V_t is the total volume flow rate of the gas mixture, A is the surface area of the sample, B is the pCO_2/pCO ratio, and $p^{14}CO$ and $(p^{14}CO)_{eq}$ are the partial pressure of tagged CO after the reaction and if complete isotope equilibrium were established. The unit of k is $mol/(cm^2\ sec\ atm)$. Details of how to deduce eq. (4-11) can be found elsewhere [3].

The evaluated rate constants k_a in the temperature range of 1823 K to 1923 K with various pCO/pCO_2 ratios are presented in the Arrhenius plot in Fig. 6-3. The S content of the samples was 0.0022 mass% [2].

The measured k_a at 1873 K (indicated by a broken line in Fig.6-3) is presented in Fig. 6-4 to show the effect of pCO/pCO_2 on the reaction rates. The CO_2/CO ratio of 6.05 when iron oxide is formed is shown by the broken line in the figure.

The apparent rate constant k_a decreases with the decrease of pCO/pCO_2. The same tendency was also observed at different temperatures. These results simply suggest that adsorbed oxygen blocks the Fe surface. Accordingly, we may write,

$$k_a = k_f (1 - \theta_0) \qquad \text{----- (5-22)'}$$

where k_a and k_f are the apparent rate constant and true forward reaction constant for the dissociation of CO_2 at the bare iron surface, respectively. θ_0 is the fractional coverage by adsorbed oxygen. The k_0 in (5-22) is replaced by k_f in (5-22)'

Since the isotope exchange reaction is carried out under the thermodynamic equilibrium condition, the activity of O(ad) can be represented by pCO_2/pCO in the following reaction

$$CO_2(g) = O(ad) + CO(g) \qquad \text{----- (6-14)}$$

Tentatively assuming that the adsorption of O(ad) will follow Langmuir adsorption isotherm, we may write

$$\theta_0/(1 - \theta_0) = K_O P_{CO2}/P_{CO} \qquad \text{----- (6-15)}$$

$$1 - \theta_0 = 1/[1 + K_O(pCO_2/PCO)] \qquad \text{----- (6-16)}$$

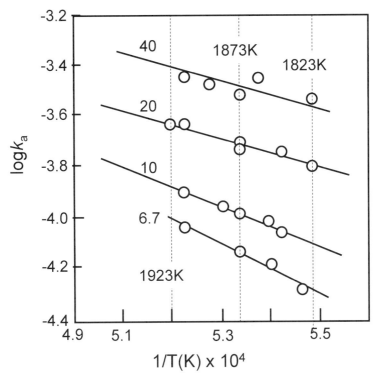

Fig. 6-3 The derived values of rate constants from the isotope exchange measurements [2]. Values of the pCO/pCO_2 ratio are indicated for each set of results. The unit of k_a is mole/(cm²sec atm)

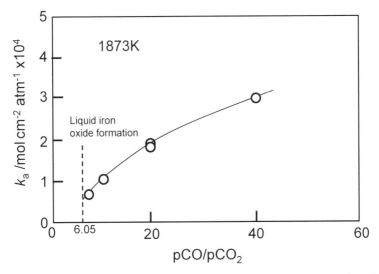

Fig. 6-4 The apparent rate constant at 1873 K as a function of pCO_2/pCO ratio [2]. The unit of k_a is mole/(cm²sec atm)

98

where K_O is the adsorption coefficient of oxygen with respect to the pCO$_2$/pCO ratio. Combined (6-13) and (6-16), we obtain

$$1/k_a = 1/k_f + K_O/k_f P_{CO_2}/P_{CO}$$ ----- (6-17)

If the behavior of O(ad) is expressed by the ideal Langmuir isotherm, $1/k_a$ and pCO$_2$/pCO have a linear relationship, and k_f and K_O can be evaluated from the intersection and the slope of the plot. The values of $1/k_a$ as a function of pCO$_2$/pCO at 1773 K are plotted in Fig. 6-5.

As shown in Fig. 6.5, there is a good linear relation between $1/k_a$ and pCO$_2$/pCO. It means that the ideal Langmuir adsorption isotherm can reasonably express the adsorbed oxygen behavior on Fe-C melts. Furthermore, as shown in Fig. 6.5, isotope exchange reactions are extrapolated smoothly to the reciprocal value of the rate constant for the decarburization reaction of Fe-C melt (carbon saturated) by the CO$_2$ obtained by Sain and Belton [1]. This is strong supporting evidence that the dissociation of CO$_2$,

$$CO_2(ad) \rightarrow O(ad) + CO(ad)$$ ----- (6-3)

is the rate-determining step in the decarburization of Fe-C melt (carbon saturated). Accordingly, the rate constant for the decarburization of Fe-C melt by CO$_2$ obtained by Sain and Belton [1] can be assumed to apply to the dissociation of CO$_2$ on an essentially oxygen-free surface, i.e., equal to k_f (presented by eq. (6-9)).

The absorption coefficient for oxygen K_O concerning pCO$_2$/pCO can be obtained from the slopes in Fig. 6.5 based on eq. (6-13) and results at other temperatures into account, the results

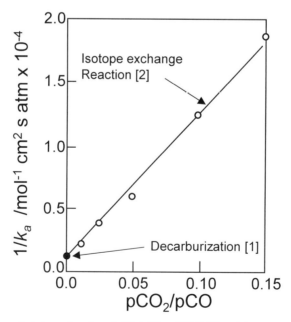

Fig. 6-5 $1/k_a$ (open circles) as a function of pCO$_2$/pCO at 1773 K [2], and the values of the reciprocal of the rate constants (closed circles) for decarburization of Fe-C melt at the bare surface by CO$_2$ obtained by Sain and Belton [1].

are expressed by

$$\log K_O = 2910/T + 0.47 \qquad \text{----- (6-18)}$$

For the following equilibrium reaction,

$$\underline{O} \text{ (dissolved in Fe)} + CO = CO_2 \qquad \text{----- (6-19)}$$

The activity of dissolved oxygen $\underline{O}$ can be expressed by

$$[\text{mass\% O}] = K_{19}(pCO_2/pCO) \qquad \text{----- (6-20)}$$

where K_{19} is the equilibrium constant for (6-19), and [mass% O] is the oxygen activity for the infinite dilute solution with 1 mass% standard state.

The adsorption coefficient with [mass% O] may be readily derived by combining (6-18) and (6-20) with appropriate thermodynamic data. The absorption coefficient for oxygen $K_O{'}$ concerning the infinitely dilute solution with 1 mass% standard state is given by

$$\log K_O{'} = 11370/T - 4.09 \qquad \text{----- (6-21)}$$

In this case, Langmuir isotherm is

$$\theta_O/1 - \theta_O = K_O[\text{mass\% O}] \qquad \text{----- (6-22)}$$

Using eq. (6-22), the adsorbed oxygen coverage θ_O at 1823 K is calculated as a function of oxygen activity expressed by oxygen mass % (a standard state for the dissolved oxygen of 1 mass% solution), and the typical results are shown in Table 6-1. $K_O{'}$ at 1823 K is about 140.

From Table 6-1, the available surface area for CO$_2$ adsorption at the surface of liquid Fe-0.1 mass% O is only about 7 % of the geometrical surface area. Even though the liquid Fe contains only 10 ppm of oxygen, about 12% of the surface is covered by adsorbed oxygen.

Using the following thermodynamic data,

$$2CO + C = CO_2 \qquad \Delta G^0 = -166.4 + 0.1707T \qquad \text{kJ/mol}$$
$$FeO(l) + CO = Fe(l) + CO_2 \qquad \Delta G^0 = -24.87 + 0.031T \qquad \text{kJ/mol}$$

the pCO$_2$/pCO ratio corresponding to the carbon saturated condition of Fe-C melt at 1873 K is calculated to 5.11 × 10^{-5}. Using eq. (6-22), the calculated θ_O at Fe-C melt with carbon saturation is 7 × 10^{-3}. Thus, the assumption of *bare surface* at the carbon saturated condition is reasonable.

The pCO$_2$/pCO ratio under liquid Fe and molten iron oxide equilibrium condition at 1873 K is calculated to be 0.110. With this pCO$_2$/pCO ratio, θ_O is about 0.939. That is, iron oxide is formed before O(ad) full coverage. After θ_O reaches a coverage of 0.939 at 1873 K, the oxygen supply rate is smaller than the removal rate. The difference in the rates results in the formation

Table 6-1   Oxygen coverage on liquid iron as a function of oxygen mass%.

mass%	θ_O
1	0.993
0.1	0.933
0.01	0.583
0.001 (10 ppm)	0.122

of iron oxide.

6.3   The effect of S adsorption

Sulfur is also known as a surface-active element and will block the reaction sites. Sain and Belton carried out decarburization experiments of Fe-C-S melts with varying sulfur contents under carbon-saturated conditions [4]. The decarburization rate was found to decrease with increasing S content, as we expect. They confirmed that the adsorption behavior of S on Fe-C-S melt (C-saturated) with low S concentration was reasonably expressed by Langmuir adsorption isotherm for a wide range of S concentrations. That is,

$$\theta_S/1 - \theta_S = K_S \text{[mass\%S]} \qquad \text{----- (6-23)}$$

where θ_S is the fraction of S coverage and K_S is the adsorption coefficient for S. As the activity scale of S, mass% S in C-saturated iron is used by assuming Henry's law for convenience. K_S was expressed by

$$\log K_S = 3600/T + 0.56 \qquad \text{----- (6-24)}$$

Although the behavior of S adsorption is well described by Langmuir adsorption isotherm, it showed a unique adsorption character for the Fe melt with high S concentration. Since the S adsorption behavior was expressed by Langmuir adsorption isotherm, the apparent decarburization rate constant k_a can be expressed by

$$k_a = k_f/(1 + K_S \text{[mass\% S]}) \qquad \text{----- (6-25)}$$

Under a high S concentration range ($1 \ll K_S$[mass%S]), eq. (6-25) is approximated to

$$k_a \approx k_f/(K_S \text{[mass\% S]}) \qquad \text{----- (6-26)}$$

where k_f is the rate constant at a bare surface. From eq. (6-26), k_a will be proportional to 1/[mass% S] and approach to zero with an increase of [mass% S] under a high S concentration range.

The decarburization rates (expressed by carbon weight-decrease, g/min) of Fe-C (carbon saturated) melts by CO_2 at 1773 K with S various concentrations more than 0.05 mass% are presented as a function of 1/[mass% S] in Fig. 6-6

The results show a good linear dependence rate on 1/[mass% S] as we expect but with a residual rate at [1/S mass%] → 0. The linear dependency means that the adsorption of S on Fe melt is possibly presented by Langmuir adsorption isotherm. The residual rate implies that some reaction sites are still available for reactions even under full S coverage.

In the same way, under the high O concentration range, the CO_2 dissociation rate on the Fe-C-S melt k_a is approximately expressed by

$$k_a \approx k_f/(K_O \text{[pCO}_2/\text{pCO]}) \qquad \text{----- (6-27)}$$

where the activity of oxygen a_O is defined by pCO_2/pCO. The CO_2 dissociation rate constants at O concentrations above 0.04 mass% are presented as a function of 1/[pCO_2/pCO] in Fig. 6-7.

The results show a good linear dependence rate on 1/(pCO_2/pCO) over a high oxygen con-

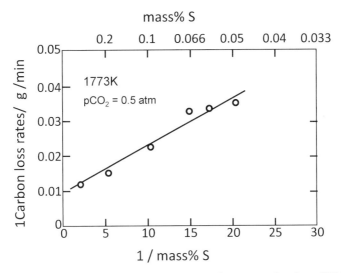

Fig. 6-6 Carbon loss during the decarburization of Fe-C (carbon saturated) melts at 1773 K as a function of the reciprocal sulfur concentration [4]. The reducing gas is CO-50% CO_2.

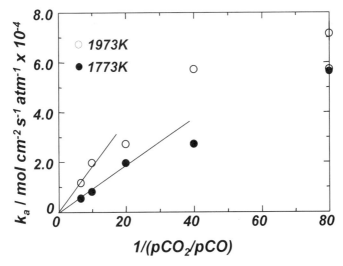

Fig. 6-7 The apparent rate constant of the CO_2 dissociation rate on Fe melt as a function of $1/(pCO_2/pCO)$.

centration range. Unlike S adsorption, k_a approaches 0 at $1/(pCO_2/pCO) \rightarrow 0$. That is, it shows ideal Langmuir adsorption behavior with no residual rate. Thus, in O adsorption, no reaction sites are available under full O coverage. However, full O_{ad} coverage never occurs since the iron oxide forms before θ_O becomes 1, as shown in Fig. 6-4.

The different adsorption behaviors of O and S can be qualitatively explained from the diameters of the adsorbed species. It is believed that the diameters of O(ad) and S(ad) are close

Fig. 6-8   Ionic diameters of sulfur and oxygen ions, and metallic Fe atom diameter.

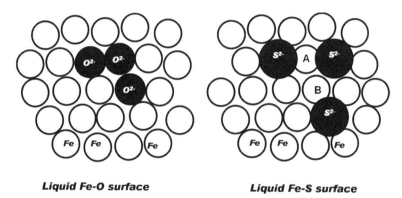

Liquid Fe-O surface   **Liquid Fe-S surface**

Fig. 6-9   Expected surface structure of the adsorbed S and O on liquid iron.

to the ionic diameters of O^{2-} and S^{2-}. The Fe diameter and ionic diameters of S^{2-} and O^{2-} are shown in Fig. 6-8

The diameters of O^{2-} and Fe are relatively close, but S^{2-} is much larger than Fe. The adsorptions of S and O on the liquid Fe surface are schematically illustrated in Fig. 6-9. Oxygen and sulphur are assumed to adsorb on the top of the Fe atom.

In the case of O adsorption, none of the adsorbed oxygen atoms prevent another oxygen adsorption at the sites next to it. However, because the adsorbed S can cover a larger area than those of O(ad), S may not adsorb at the surrounding sites of the already adsorbed S, such as at points A or B shown in Fig. 6.9. This unavailability is one reason for the residual reaction rate for the CO_2 dissociation rate on Fe-C-S (C-saturated) melt.

Further reading

E. T. Turkdogan: Physical Chemistry of High Temperature Technology. Academic Press, New York, 1980.

A. Ozaki: Isotope Studies of Heterogeneous Catalysis, Academic Press New York 1977.

References

[1] D. Sain and G. R. Belton: Met. Trans. B, 7B (1976), 235-244.

[2] A. W. Cramb and G. R. Belton: Met. Trans. B, 12B (1981), 699-704.

[3] A. Ozaki: Isotope Studies of Heterogeneous Catalysis, Kodansha Ltd., Tokyo and Academic Press Inc., New York, NY, 1977, 25-27.

[4] D. Sain and G. R. Belton: Met. Trans. B, 9B (1978), 403-407.

Chapter 7
The reaction of CO with molten Fe (Carburization of Fe by CO)

Chapter 7   The reaction of CO with molten Fe (Carburization of Fe by CO)

7.1   CO dissociation rates

The carburization reaction of iron by CO-bearing gas mixtures occurs in the ironmaking process. Many studies of this reaction have been carried out due to its importance. The following equation expresses the carburization reaction of liquid Fe by CO gas.

$$2CO \rightarrow CO_2 + \underline{C} \qquad \text{----- (7-1)}$$

$\underline{C}$ is the dissolved carbon in Fe. The apparent reaction (7-1) consists of several elementary steps.

$$CO(g) \rightarrow CO(ad) \qquad \text{----- (7-2)}$$
$$CO(ad) \rightarrow C(ad) + O(ad) \qquad \text{----- (7-3)}$$
$$C(ad) \rightarrow \underline{C} \qquad \text{----- (7-4)}$$
$$O(ad) \rightarrow \underline{O} \qquad \text{----- (7-5)}$$
$$O(ad) + CO(ad) \rightarrow CO_2(ad) \qquad \text{----- (7-6)}$$
$$CO_2(ad) \rightarrow CO_2(g) \qquad \text{----- (7-7)}$$

(g) refers to gas, (ad) refers to the adsorbed species on Fe surface. $\underline{O}$ is the dissolved oxygen in Fe. In this way, the overall reaction occurs via a series of elementary reactions involving adsorbed species that combine to give the overall reaction (7-1).

It is well known that the CO dissociation rate on Fe is very fast, and that it occurs readily on the iron surface even at low temperatures of 195 K. [1] The CO decomposition mechanism on Fe has been well explained based on the molecular orbital simulation. Details will be discussed in Section 7.3.

In the reaction of carburization with CO-H_2 (CO + H_2 → $\underline{C}$ + H_2O), the following O(ad) removal reaction

$$O(ad) + H_2(ad) \rightarrow H_2O(ad) \qquad \text{----- (7-8)}$$

occurs instead of the reaction (7-6). The reaction (7-8) also consists of several reaction steps with H and OH species. However, the reactions involving H and OH species do not affect the frame of the following discussion and these reactions are not shown here. The carburization rates of solid Fe by CO-H_2 are much faster than those by CO-CO_2 [2-4]. This result means that the rate-controlling step of the carburization of Fe with CO or CO-H_2 are the reaction (7-6) or (7-8).

Namely, the carburization rates are controlled by O(ad) removal rates. In other words, the rates of the adsorption of CO (reaction (7-2)) and the CO dissociation reaction (reaction (7-3)) is much faster than the reactions of (7-6) and (7-8). If so, the reaction (7-3) can be assumed to be essentially under the equilibrium condition during the carburization reaction. Consequently, the CO concentration practically does not change during the carburization reaction except at the very early stage since the reaction (7-3) quickly reaches equilibrium.

Itoh *et al.* measured C and O transfer rates between CO-CO_2 gas mixture and a levitated molten iron droplet at 2073 ~ 2373 K [5]. Their experimental setup is shown in Fig. 7-1. After a particular reaction time (2 sec to 2 min), the power switch was off, and a levitated droplet was dropped to a water-cooled Cu mold for quenching.

The change of $\underline{C}$ and $\underline{O}$ content during the reaction at 2098 K is shown in Fig. 7-2. The initial composition of molten Fe was at point A (0.002 mass% C). Within about 3 seconds, the concentrations of $\underline{C}$ and $\underline{O}$ moved from A to point B. After reaching point B, they moved slowly towards point C or D along the curve DBC depending on the introducing gas mixture's CO_2/CO ratio. The curve DBC is the $\underline{C}$ - $\underline{O}$ curve equilibrated with the reaction of (7-3) under $CO + CO = 1$ atm. The slope of the line AB was about 4/3 and is corresponded to the atomic mass ratio of oxygen and carbon. This means the CO dissociation rate in the early stage (A to B) was dominantly prevailing, and the $\underline{O}$ removal rate of the reaction (7-6) was negligibly low in the early stage (A to B). It is noted that the reactions (7-4) and (7-5) are so fast that we can reasonably assume they are in virtual equilibrium.

$$C(ad) = \underline{C} \qquad\qquad ----- (7\text{-}4)'$$
$$O(ad) = \underline{O} \qquad\qquad ----- (7\text{-}5)'$$

Thus, the removal rate of O(ad) is the same as that of $\underline{O}$.

The A to B reaction can be regarded as a single CO gas absorption system or to have no CO gas concentration gradient towards the molten iron surface when pure CO gas was used. Consequently, the gas phase mass transfer in the early stage can be neglected. However, the liquid phase mass transfer cannot be overlooked.

Using their experimental parameters and neglecting the sulfur blocking effect, the estimat-

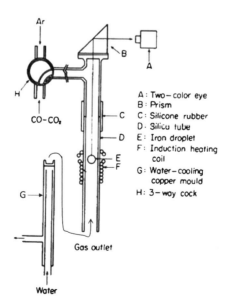

A : Two-color eye
B : Prism
C : Silicone rubber
D · Silica tube
E : Iron droplet
F : Induction heating
    coil
G : Water-cooling
    copper mould
H : 3-way cock

Fig. 7-1 Experimental apparatus for the measurement of CO dissociation rates [5].

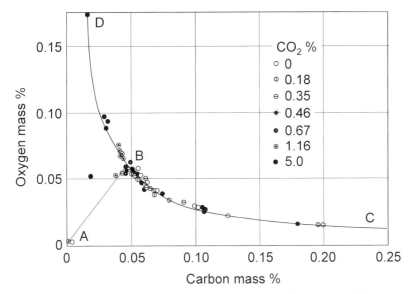

Fig. 7-2 $\underline{C}$ and $\underline{O}$ content in liquid Fe with the reaction of CO at 2098 K [5].

ed CO dissociation reaction rate at 2073 K was at least about 1.0×10^{-2} mol/cm²sec, even containing the liquid phase mass transfer resistance. Note that the estimated reaction rate constant of CO_2 dissociation at 2073 K is about 2.1×10^{-3} mol/cm²sec, as calculated from eq. (6-9).

Negative temperature dependency

The reaction of (7-3) was too fast to measure the accurate rates. Consequently, reliable activation energy was not obtained. Intriguingly, however, the range B to C (shown in Fig. 7-2) was found to decrease with temperature increase [5]. In what cases are the negative activation energy possibly observed?

The gas-phase mass transfer process may have controlled the carburization reaction in B to C (Fig. 7-2). Then, all of the elementary reactions that consisted of the carburization reactions are under equilibrium. The following reaction (7-9) is also under the equilibrium condition.

$$CO(g) + \underline{O} = CO_2(g) \qquad \text{----- (7-9)}$$

In their experiments, the gas-phase mass transport process controls the removal rate of $\underline{O}$. In this case, the mass transfer rate R_{mt} can be expressed by

$$R_{mt} = m \, (*pCO_2 - pCO_2) \qquad \text{----- (7-10)}$$

where m is the mass transfer coefficient and $*pCO_2$ is the partial equilibrium pressure of CO_2 in the reaction (7-9). pCO_2 is the partial pressure of CO_2 outside of the boundary layer. In their experiments, pCO_2 can be assumed to be zero and eq. (7-10) is

$$R_{mt} \cong m \, *pCO_2 \qquad \text{----- (7-10)'}$$

m is represented by

$$m = D/\delta \qquad\qquad ----- (7\text{-}11)$$

where D is the diffusion coefficient of gas mixture and δ is the boundary layer thickness. Generally, D and δ do not change so much with temperatures, or m is very weakly dependent on temperature, and practically almost constant with temperature. Fig. 7-3 shows the equilibrium relationship among mass% $\underline{O}$, mass% $\underline{C}$, and CO_2 of the reaction (7-9) [6].

As shown in Fig. 7.3, $*pCO_2$ in the reaction (7-9) decreases with the increase in temperature at any $\underline{C}$ concentration. If so, R_{mt} decreases with temperature since the value of $*pCO_2$ in (7-10)' decreases with the increase in temperature. Therefore, the negative temperature dependency of the carburization reaction in the range B to C is due to the decrease in pCO_2* with the temperature, based on eq. (7-10)'.

7.2 Isotope exchange reaction of CO dissociation

Due to the extremely fast reaction rate of the CO dissociation at high temperatures, it is hard to carry out experiments without the influence of the gas and liquid phase mass transfer resistance. One method to eliminate or minimize the effect of the gas and liquid phase mass transfer resistance is to use the isotope exchange reaction of

$$^{13}C^{18}O + {}^{12}C^{16}O + \rightarrow {}^{13}C^{16}O + {}^{12}C^{18}O \qquad ----- (7\text{-}12)$$

For the CO dissociation reaction rate measurements by applying the isotope exchange reaction, it is necessary to use a "double-labeled isotope" of CO consisting of ^{13}C and ^{18}O ($^{13}C\,^{18}O$).

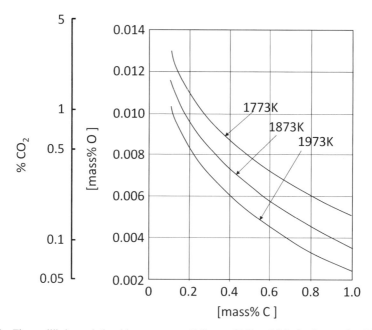

Fig. 7-3 The equilibrium relationship among mass% $\underline{O}$, mass% $\underline{C}$, and CO_2 for the reaction $CO(g) + \underline{O}$ → $CO_2(g)$ under the condition of $pCO_2 + pCO = 1$.

If a single-labeled isotope $^{13}C\ ^{16}O$ or $^{12}C\ ^{18}O$ is used, there will be no net change in the isotope concentration during the exchange reactions.

The $^{13}C^{18}O$ dissociation reaction would be

$$^{13}C^{18}O \rightarrow\ ^{13}C + ^{18}O \qquad\qquad ----- (7\text{-}13)$$

For the exchange reaction of (7-12), the introducing gas mixture generally consists of a large amount of $^{12}CO^{16}O$ and a small amount of $^{13}C^{18}O$. Consequently, much more ^{16}O and ^{12}C are adsorbed at the surface than ^{18}O and ^{13}C. In other words, the chance to recombine ^{13}C and ^{18}O atoms after $^{13}C^{18}O$ dissociation is negligibly small, and almost all of ^{13}C will react with ^{16}O and almost all of ^{18}O will react with ^{12}C, respectively, to form $^{13}C^{16}O$ and $^{12}C^{18}O$.

$$^{13}C + ^{16}O \rightarrow\ ^{13}C^{16}O \qquad\qquad ----- (7\text{-}14)$$
$$^{12}C + ^{18}O \rightarrow\ ^{12}C^{18}O \qquad\qquad ----- (7\text{-}15)$$

Namely, the recombination of ^{13}C and ^{18}O at the Fe surface is negligible. Therefore, the rate of the CO dissociation can be evaluated from the measurements of the $^{13}C^{18}O$ concentration changes. However, if we were to use a single-labeled isotope, such as $^{13}C^{16}O$, the dissociation reaction would be

$$^{13}C^{16}O \rightarrow\ ^{13}C + ^{16}O \qquad\qquad ----- (7\text{-}16)$$

The produced ^{13}C can react only with ^{16}O to form $^{13}C^{16}O$,

$$^{13}C + ^{16}O \rightarrow\ ^{13}C^{16}O \qquad\qquad ----- (7\text{-}14)$$

Consequently, the overall exchange reaction is expressed by

$$^{13}C^{16}O + ^{12}C^{16}O + \rightarrow\ ^{13}C^{16}O + ^{12}C^{16}O \qquad\qquad ----- (7\text{-}17)$$

Namely, there is no net change in $^{13}C^{16}O$ concentration.

In Chapter 6, the CO_2 dissociation reaction rates were studied by applying the following isotope exchange reaction.

$$^{14}CO_2 + ^{12}CO \rightarrow\ ^{14}CO + ^{12}CO_2 \qquad\qquad ----- (4\text{-}6)$$

During this exchange reaction, ^{14}CO dissociation is also expected to occur. However, ^{14}CO is a single-labelled isotope; there is no net change in the $^{14}C^{16}O$ concentration during the reaction. Thus, it is possible to neglect the ^{14}CO dissociation in (4-6) exchange reaction.

In the CO dissociation rate measurements, eliminating the gas phase and liquid phase mass transfer resistances is quite difficult since the CO dissociation rate is extremely fast. The isotope exchange reactions are carried out under the equilibrium condition, the liquid phase mass transfer contribution to the reaction rate can be automatically avoided.

In the isotope exchange reaction of $^{13}C^{18}O + ^{12}C^{16}O \rightarrow\ ^{13}C^{16}O + ^{12}C^{18}O$, there is no apparent concentration gradient of CO in the vicinity of the interface since the difference of diffusivities among these CO gases is essentially negligible due to their small mass differences among CO gases. Therefore, only the back diffusion of $^{13}C^{16}O$ and $^{12}C^{18}O$ from the reacting surface should be avoided to eliminate the gas phase mass transfer resistance. In other words, a small gas flow rate is sufficient to eliminate the back diffusion in general.

Fruehan and Antolin tried to measure CO dissociation rates on liquid Fe by using the isotope exchange reaction of $^{13}C^{18}O + {}^{12}C^{16}O \rightarrow {}^{13}C^{16}O + {}^{12}C^{18}O$ [7]. They found that the rates did not change significantly with temperature, sulfur activity, or carbon content but increased with the gas flow rates. These results indicate that gas-phase mass transfer may still influence the overall rate of the reaction even with the total flow rate of 10 l/min, or the dissociation rate of CO was confirmed to be extremely fast. Most of the experiments were controlled by gas-phase mass transfer, but rates with high S content Fe were likely controlled by mixed control. The estimated chemical rate constant at 1523 K under this condition of high S content was about 1.5×10^{-5} moles/cm^{2}sec. The rates decreased with the increase of S under mixed control regime. Thus, S is reasonably assumed to block the CO adsorption sites. Considering Langmuir adsorption isotherm for S, the coverage of S with 0.1 mass% at 1523 K is about 99% by using eq. (6-22). Under this condition, the expected rate constant at the bare surface will be about 1.87×10^{-3} moles/cm2sec. The rate constant of CO_2 dissociation at 1523 K is about 2.8×10^{-4} moles/cm^{2}sec.

7.3 Molecular orbital approach to the CO dissociation reaction mechanism

The CO dissociation reaction occurs very selectively, or it occurs only at the surface of certain metals. For example, it occurs at the surface of Fe, Ni, and Cr, but it does not occur at the surface of Au, Ag, or Pt. The reason for this selectivity is well explained based on quantum chemistry by Sung and Hoffmann [8].

The electron energy levels of CO gas and metallic Fe are shown in Fig. 7-4. For the molecular orbital of CO, only the antibonding orbital of $2\pi^*$ and bonding orbital 5σ are shown. Two electrons fully occupy the bonding orbital of 5σ, but the anti-bonding orbital of $2\pi^*$ is not occupied. The highest energy levels in metals are located near the Fermi energy (E_F) level, and only these electrons can interact with other materials. As shown in Fig. 7-4, The energy level of E_F in Fe is much higher than the unoccupied antibonding $2\pi^*$ level in CO. Therefore, when CO gas approaches the Fe surface, the electrons in the vicinity of E_F can move to $2\pi^*$ orbital.

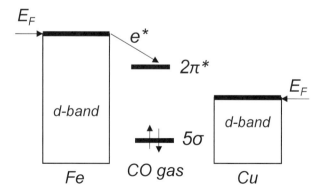

Fig. 7-4 The electron energy levels of CO gas and metallic Fe.

Table 7-1   Some orbital population in CO chemisorption on first transition series surface.

	Electron densities in fragment orbitals					
	Ti (0001)	Cr (110)	Fe (110)	Co (0001)	Ni (100)	Ni (111)
5σ	1.73	1.67	1.62	1.60	1.60	1.59
$2\pi^*$	1.61	0.74	0.54	0.43	0.39	0.40

When electrons exist in the $2\pi^*$ anti-bonding orbital, the possibility of CO dissociation becomes high when the vibration between the C and O atoms is larger.

E_F in Cu is lower than the $2\pi^*$ energy level, so electrons at E_F in Cu cannot move to $2\pi^*$ orbital. Thus, there is no chance for CO to dissociate at the Cu surface.

The CO dissociation mechanism at the Fe surface is a bit more complex. However, the fundamental feature of CO dissociation at the Fe surface that is driven by $2\pi^*$ orbital occupation by the electrons at the Fermi energy level of Fe is not changed.

Early and middle transition metals can generally break up CO. Sung and Hoffman calculated the 5σ and $2\pi^*$ orbital population in CO chemisorbed on the first transition series by using Extended Hückel method [8]. The results are shown in Table 1. The population of 5σ slightly increases as one moves from Ni to Ti. The population of $2\pi^*$, however, increases sharply. If CO were allowed to stretch by thermal activation, CO would surely and quickly dissociate on the left side of the series.

7.4   Carburization rate of high carbon contained Fe-C melt by CO

The schematic diagram of the equilibrium relationship between mass% $\underline{O}$ and mass% $\underline{C}$ in the liquid Fe-C system is shown in Fig. 7-5.

As explained in Section 7.1, Fe carburization is carried out by CO dissociation reaction (7-3) in the early stage (A to B in Fig. 7-5). After it reaches B, when the oxygen removal rate is faster than the oxygen supply rate, carburization proceeds along the BC line. In the reverse case, decarburization occurs along the line BD. In this section, the carburization reaction rate in the range B to C has been evaluated. In the range B to C, the reactions of (7-18) and (7-19) occur simultaneously.

$$CO(g) \rightarrow \underline{C} + \underline{O} \qquad \text{----- (7-18)}$$
$$\underline{O} + CO(g) \rightarrow CO_2(g) \qquad \text{----- (7-19)}$$

In this range, the reaction (7-18) provides $\underline{C}$ and $\underline{O}$, and the reaction (7-19) removes $\underline{O}$. Since the CO dissociation rate (7-18) is under equilibrium, $\underline{C}$ gradually increases with the decrease of $\underline{O}$ along with the decrease of carbon content. When the gas and liquid phase mass transfer resistances are negligibly small, the rate-controlling step during the carburization reaction from B to C is the reaction (7-19).

Thus, the overall carburization rate in the range B to C is expressed by

$$d[C]/dt = k_2 pCO\theta_0 \qquad \text{----- (7-20)}$$

114

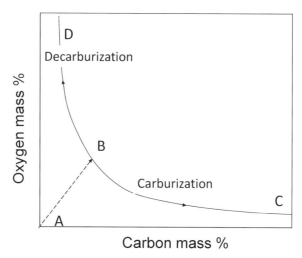

Fig. 7-5 The schematic diagram of the equilibrium relationship between mass% $\underline{O}$ and mass% $\underline{C}$ in the liquid Fe-C system.

where k_2 is the rate constant (mole/cm^2 s atm) of the reaction (7-19), [C] is carbon concentration (mole%). To evaluate the d[C]/dt, we need to know k_2 and θ_0. k_2 can be calculated from the isotope exchange reaction of (4-6),

$$^{14}CO_2 + {}^{12}CO \rightarrow {}^{14}CO + {}^{12}CO_2 \qquad \text{----- (4-6)},$$

The rate-controlling reaction of (4-6) was

$$^{14}CO_2(ad) \rightarrow {}^{14}CO(ad) + O(ad) \qquad \text{----- (4-7)}.$$

Since the exchange reaction (4-6) was carried out under equilibrium conditions, the rate of forwarding reaction (4-7) is the same as that of the backward reaction (7-21).

$$^{14}CO(ad) + O(ad) \rightarrow {}^{14}CO_2 \qquad \text{----- (7-21)}$$

The rate of (4-7) can be expressed by $k_1 p^{14}CO_2(1-\theta_0)$ and that of (7-21) by $k_2 p^{14}CO\theta_0$. k_1 and k_2 are forward and backward reaction rate constants, respectively. Since both of the rates are equal under equilibrium conditions,

$$k_1 p^{14}CO_2(1 - \theta_0) = k_2 p^{14}CO\theta_0 \qquad \text{----- (7-22)}$$

θ_0 is readily calculated by Langmuir adsorption isotherm once the oxygen activity is known,

$$\theta_0/(1 - \theta_0) = K_O[O \text{ mass}\%] \qquad \text{----- (7-23)}$$

where K_O is the adsorption constant and [O mass%] is the oxygen activity for the infinitely dilute solution with 1 mass% standard state. Combined (7-22) and (7-23),

$$k_2 = k_1/K_O \qquad \text{------ (7-24)}$$

As already mentioned in 4.3.1, the (4-7) and (7-21) rates are the same as those untagged ones. Thus, the relation (7-24) can be applied for the untagged reactions.

k_1 and K_O are already obtained by

$$\log k_1 = -5080/T - 0.21 \qquad \text{----- (6-9)}$$

$$\log K_O = 2910/T + 0.47 \qquad \text{----- (6-18)}$$

From eq. (7-24), (6-9) and (6-18),

$$\log k_2 = -7990/T - 0.68 \qquad \text{----- (7-25)}$$

In the case of high carbon melt, $\theta_0 \approx K_O[\text{O mass\%}]$. Then,

$$d[C]/dt = k_2 pCO\theta_0 \approx k_2 K_O \, pCO[\text{mass\% O}] \qquad \text{----- (7-26)}$$

For example, the decarburization rate of Fe-C melt ($\underline{O} = 0.006$ mass%, $\underline{C} = 0.4$ mass%) at 1873 K is calculated to 8.83×10^{-7} (mol/cm^2 s atm). It is noted that the carburization rate in high carbon Fe-C melts decreases with a decrease of [mass% O], as shown in eq. (7-26).

7.5 Decarburization rate of low carbon contained Fe-C melt by CO-CO$_2$ gas mixtures

Recent development of some fine steel demands the carbon content as minimum as possible.

However, a few studies of the decarburization for low carbon contained Fe-C melts have been reported [9, 10]. As shown in Fig. 7-5, when the oxygen supply rate is larger than the oxygen removal rate, the decarburization proceeds along the line BD in Fig. 7-5. In this range, the carbon content is relatively small, and the oxygen is supplied by the reaction (7-18), and the carbon content decreases with the increase of oxygen by the reaction (7-27).

$$CO_2(g) \rightarrow CO(g) + \underline{O} \qquad \text{----- (7-27)}$$

Nomura *et al.* [10] studied the decarburization of a liquid Fe-C melt with a low carbon content of less than 0.1 mass%. The rate of carbon loss as a function of carbon mass% for varying pCO_2/pCO is shown in Fig. 7-6, and that rate at a carbon concentration of 0.04 mass% as a function of pCO_2/pCO is shown in Fig. 7-6.

The rates as a function of pCO_2/pCO at the carbon concentration of 0.04 mass% derived

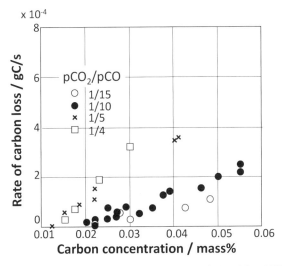

Fig. 7-6 The carbon loss rates in liquid iron with the reaction with CO$_2$-CO at 1873 K as a function of carbon concentration with various pCO_2/pCO ratios from the results of Nomura et al. [11].

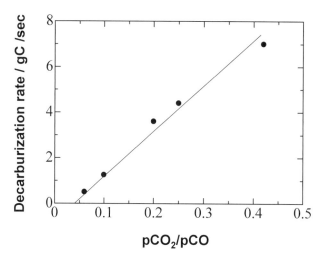

Fig. 7-7   The carbon loss rates in liquid iron at the carbon concentration of 0.04 mass% with CO_2-CO at 1873 K as a function of with pCO_2/pCO ratios [10].

from Fig. 7-7 give a straight line. That is, the rate can be expressed by

$$\text{Rate} = k_{ap}A(pCO_2/pCO)\,[C] \qquad ----- (7\text{-}28)$$

where k_{ap} is the apparent rate constant, A is the reacting surface area, and [C] is the concentration of carbon in mass%. The rate is proportional to the pCO_2/pCO ratio and [C]]. The rate law (7-28) differs from (6-8) of carbon saturated conditions.

$$d[C]/dt = -k_f pCO_2 \qquad ----- (6\text{-}8)$$

They also found that the variation of carbon and oxygen concentrations during the decarburization process corresponded to the equilibrium value of the reaction (7-18)

$$CO(g) \rightarrow \underline{C} + \underline{O} \qquad ----- (7\text{-}18)$$

Accordingly, it can be assumed that the reaction (7-18) is at virtual equilibrium during the decarburization process, and the CO_2 dissociation on the surface of iron is the rate-limiting step. In this case, adapting the Langmuir adsorption model, the rate can be expressed by

$$\text{Rate} = d[C]/dt = -k_p ApCO_2(1 - \theta_0) \qquad ----- (7\text{-}29)$$

$(1-\theta_0)$ can be calculated using the Langmuir isotherm.

$$\theta_0/(1 - \theta_0) = K_O[\text{mass\% O}] \qquad ----- (6\text{-}22)$$

where Ko is the adsorption coefficient for oxygen at molten iron and [maass% O] is the oxygen activity, using 1 mass% oxygen as the standard state. K_O is expressed by

$$\log(K_O) = 11370/T - 4.09 \qquad ----- (7\text{-}30)$$

Combination (7-29) and (6-12) gives,

$$\text{Rate} = k_p ApCO_2/(1 + K_O[\text{mass\% O}]) \qquad ----- (7\text{-}31)$$

In their experimental conditions, the oxygen concentration varies from 0.05 to 0.1 mass%. Then, Ko[mass% O] at 1873 K has a relatively large value of about "10" compared with "1" in the denominator of Eq. (14). This reduces Eq. (7-31) to Eq. (7-32).

$$\text{Rate} = k_p A p CO_2 / K_O [\text{mass\% O}]) \qquad \text{----- (7-32)}$$

As previously mentioned, the reaction (7-18) is supposed to be at equilibrium, then

$$[\text{mass\% O}] = p CO / Kco \ Ko(aC) \qquad \text{----- (7-33)}$$

where Kco is the equilibrium constant of the reaction (7-18) and expressed by

$$\log(Kco) = 1\ 160/T + 2.003 \qquad \text{-----(7-34)}$$

aC is the carbon activity, using I mass% carbon as the standard state and is equal to [mass% C] at a low carbon concentration of below 0.05 mass%. Then,

$$[\text{mass\% O}] = p CO / Kco\ [\text{mass\% C}] \qquad \text{----- (7-35)}$$

Combining (7-32) and (7-35),

$$\text{Rate} = (k p Kco / K_O) A (p CO_2 / p CO)\ [\text{mass\% C}] \qquad \text{-----(7-36)}$$

Replacing $(k p Kco / K_O)$ by k_{app}, then

$$\text{Rate} = k_{app} A (p CO_2 / p CO\ )\ [\text{mass\% C}] \qquad \text{-----(7-37)}$$

The derived Eq. (7-37) has the same form as (7-28).

This good agreement is strong evidence that the rate-determining step of the decarburization at a low carbon concentration of below 0.05 mass% is the CO_2 dissociation reaction at the liquid Fe surface partially covered by oxygen. The different form of the rate equation between (6-8) and (7-37) is due to the variation of the term $(1-\theta_0)$ with time. At high carbon concentrations, the value of $(1-\theta_0)$ is almost equal to 1 during the carburization process since the oxygen coverage is almost negligible ($\theta_0 = 0$). However, a slight decrease in carbon in the low carbon concentration range easily produces a large increase in the oxygen content in molten iron since reaction (7-18) is fast and at virtual equilibrium. This increase of the oxygen coverage θ_0 results in a decrease in the decarburization rate.

7.6 Oxidation of CO with O_2 gas

The following reaction expresses the oxidation of CO gas by O_2.

$$2CO(g) + O_2(g) \rightarrow 2CO_2(g) \qquad \text{----- (7-38)}$$

CO gas oxidation (7-38) can be carried out as a homogeneous or a heterogeneous reaction.

Homogeneous CO gas oxidation is called CO gas combustion. It is well known that dry CO is slow to oxidize and even extremely difficult to ignite. However, the CO oxidation rate in the presence of hydrogen-containing species such as H_2O or H_2 is substantially faster than that in dry conditions. Brokaw [11] found that minimal quantities of H_2, even of about 20 ppm, considerably increases the CO oxidation rate. The addition of hydrogen-containing species introduces very reactive radicals to enhance CO oxidation.

Simply introducing CO-O_2 gas mixture into a gas reservoir at a high temperature does not result in the CO oxidation reaction. However, if a piece of Pt foil exists in the reservoir, this reaction will proceed quickly at the Pt surface even at room temperature. This heterogeneous CO gas oxidation reaction occurs daily. When driving a car, the hazardous CO gas in the exhaust gas is continuously converted to harmless CO_2 through the CO oxidation reaction on Pt catalysts in the exhaust gas converter.

In the heterogeneous CO oxidation reaction, it proceeds with the combination of the following consequent reactions.

$$O_2(g) \rightarrow 2O(ad) \qquad\qquad ----- (7\text{-}39)$$
$$CO(g) \rightarrow CO(ad) \qquad\qquad ----- (7\text{-}40)$$
$$O(ad) + CO(ad) \rightarrow CO_2(ad) \qquad\qquad ----- (7\text{-}41)$$
$$CO_2(ad) \rightarrow CO_2(g) \qquad\qquad ----- (7\text{-}42)$$

Oxygen undergoes dissociative adsorption to form two adsorbed oxygen atoms, and CO is adsorbed on the Pt surface. These two species react and form $CO_2(ad)$, and it will leave from Pt surface as $CO_2(g)$. This type of reaction (a reaction between two adsorbed species) is called the Langmuir-Hinshelwood reaction mechanism.

The Heterogeneous CO gas oxidation with O_2 only occurs at the surface of specified metals. To promote heterogeneous CO gas oxidation with O_2, CO should not dissociate at a metal surface, and O_2 must dissociate to $2O$. The CO oxidation reaction does not occur at Fe since CO adsorbs as dissociative adsorption. The reaction does not occur at Au and Ag since O_2 dissociation does not occur. It occurs on Pt but not Ni, even though they belong to the same group in the periodic table and their chemical properties are quite similar. It is confirmed that the catalytic effect of the Pt surface for CO oxidation is due to the rearrangement of the surface Pt atoms when CO gas is absorbed. On the Ni surface, this rearrangement does not occur. Although catalysis research is considered as a monopoly of chemistry, approaching surface science from a physics perspective is necessary to elucidate the mechanism.

Further reading

E. T. Turkdogan: Physical Chemistry of High Temperature Technology. Academic Press, New York, 1980.

R. Hoffmann, Solids and Surfaces – A Chemist's View of Bonding in Extended Structures, VCH Publishers Inc., 1988.

A. Ozaki: Isotope Studies of Heterogeneous Catalysis, Academic Press New York 1977

References

[1] A.N. Webb and R.P. Eischens: J. Amer. Chem. Soc., **77** (1955), 4710-4712.

[2] H. J. Grabke: Arch. Eisenhüttenwes., **46** (1975), 75-81.

[3] D. R. Stransky and H. J. Grabke: Eisenhüttenwes., **49** (1978), 9633.

[4] J. H. Kaspersma and R. H. Shay: Metall. Trans. B: **12B** (1981), 77-83.

[5] K. Itoh, K. Amano and H. Sakao: Tetsu to Hagané, **61** (1975), 312-320.

[6] The steelmaking Data source book: The 3rd Subcommittee on Steelmaking Reactions in the 19th Committee on Steelmaking, Japan Society for the Promotion of Science, Gordon and Breach Sci. Pub., New York, 1988.

[7] Fruehan and Antolin: Met. Trans B., **18B** (1987), 415-420.

[8] S-S. Sung and R. Hoffmann: J. Am. Chem. Soc., **107** (1985), 578-584.

[9] Y, Sasaki and K. Ishii: ISIJ Int., **37** (1997), 1037-1039.

[10] H. Nomura and K Mori: Tran Iron Steel Inst. Jpn., **13** (1973), 325-332.

[11] R. S. Brokaw: Proc. Combust. Inst., **11** (1967) 1063-1073.

Chapter 8
Reaction of molten Iron oxides with CO_2-CO mixtures

Chapter 8   Reaction of molten Iron oxides with CO$_2$-CO gas mixtures

8.1   Introduction

Many studies on iron oxide reduction have been carried out due to its importance in iron-making and steelmaking processes. Most of the molten iron oxides reduction experiments have used graphite crucibles (solid) or Fe-C melt (liquid) for reduction since molten iron oxides readily react with refractories. In these cases, graphite or the Fe-C melt reacts with the molten iron oxide and produces CO gas bubbles at the reaction interface. The CO gas bubble formation makes it difficult to precisely and reliably determine reaction area. Thus, only qualitative results have been achieved. In this chapter, the focus is on the reduction reactions of liquid iron oxides with CO-CO$_2$ gas mixtures, although it is acknowledged that little research has been published on the gas reduction of molten iron oxide.

The reduction of molten iron oxide can be classified into two types. One is reduction without Fe formation, and the second one is that with Fe formation. In the first type, the Fe^{3+}/Fe^{2+} ratio changes along with the reduction, whereas this ratio remains constant during the reduction in the second type. Fig. 8-1 shows these two types of reduction paths in the Fe-O phase diagram. Type I: The molten iron oxide at point A will be reduced to point B without forming Fe, and the Fe^{3+}/Fe^{2+} ratio continuously decreases until reaching point B. Type II: After reaching B, Fe formation starts, and the molten iron oxide reduction continues until all of the iron oxide are converted to Fe. During the type II reduction process, the Fe^{3+}/Fe^{2+} ratio in the molten iron oxide does not change.

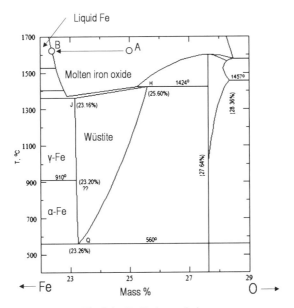

Fig. 8-1   Fe-O phase relation.

As shown in Fig. 8-1, the Fe/O ratio in solid and liquid iron oxides is always less than 1. Since there is less iron in iron oxide than oxygen, liquid and solid iron oxides always contain vacancies of holes or Fe^{3+} ions to maintain electronic neutrality. Therefore, even when Fe is in the saturated condition, the iron oxide contains Fe^{3+} ions, and the Fe^{3+}/Fe^{2+} ratio in liquid iron oxide does not change with the iron-saturated condition.

The reduction of molten iron oxide by CO gas without Fe formation (A to B in Fig. 8-1) can be expressed by

$$CO + O^{2-} \rightarrow CO_2 + 2e^- \qquad \text{----- (8-1)}$$
$$Fe^{3+} + e^- \rightarrow Fe^{2+} \qquad \text{----- (8-2)}$$

where 'e' represents electrons. Along with the oxygen (O^{2-}) removal from the liquid iron oxide surface, the ratio Fe^{3+}/Fe^{2+} decreases to maintain electrical neutrality until Fe finally appears. Since the electron conductivity in molten iron oxide is so large, the reaction of (8-1) and (8-2) can proceed at different reaction sites. The released electrons from O^{2-} can move rapidly to the remote reaction sites of the reaction (8-2).

The reduction by CO gas under the Fe saturated condition is expressed by

$$CO + O^{2-} \rightarrow CO_2 + 2e^- \qquad \text{----- (8-1)}$$
$$Fe^{2+} + 2e^- \rightarrow Fe \qquad \text{----- (8-3)}$$

For the same reason, the reaction of (8-1) and (8-3) can also proceed at different reaction sites. The Fe^{3+}/Fe^{2+} ratio in molten iron oxide does not change during the reduction process in the Fe saturated condition. It is noted that a small number of Fe^{3+} cations exist even under Fe saturated conditions.

The ratio of Fe^{3+}/Fe^{2+} in the pure molten iron oxide as a function of the CO_2/CO ratio was measured by Darken Gurry [1], and the results are presented in Fig. 8-2. There is an essential linear relationship, as follows:

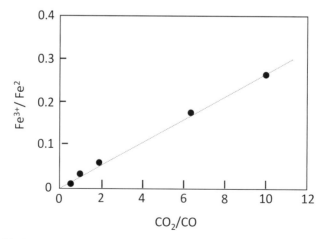

Fig. 8-2 Fe^{3+}/Fe^{2+} ratio in the pure molten iron oxide as a function of the CO_2/CO ratio at 1773 K [1].

$$(Fe^{3+}/Fe^{2+})^2 = 0.027 \ (pCO_2/pCO). \qquad\qquad ----- (8\text{-}4)$$

They found that there was little change in the data at 1673 K and 1873 K. This suggests that the relation (8-4) is practically independent of temperature.

The ratio pCO_2/pCO with the Fe saturated condition at 1673 K is about 0.261. Using eq. (8-4), the ratio Fe^{3+}/Fe^{2+} at this condition is about 0.084. In other words, 7.7% of Fe^{3+} still exists even under Fe saturated conditions. Since the reduction of the molten iron oxide involves the transfer of electrons, the change of the Fe^{3+}/Fe^{2+} ratio certainly influences the overall molten iron oxide reduction rate. In other words, to understand the reaction mechanism, the effect of the ratio Fe^{3+}/Fe^{2+} must be considered. Once Fe is formed, the Fe^{3+}/Fe^{2+} ratio does not change, and the ratio does not affect the rate during the reduction.

8.2   Reduction of molten iron oxide under the Fe saturated condition

8.2.1   Reduction of molten iron oxide droplets in a gas conveyed system

Gas conveyed reacting technology is defined as the direct gaseous reduction of fine ore concentrates transported by reducing gas. The process uses several gaseous reducing agents, such as CO, hydrogen, and natural gas. This technology has been widely used in many metallurgical fields as "flash," "cyclone," and "in-flight" processes. The gas conveyed system has several advantages for the gas reduction of molten iron oxide. (1) Molten iron oxide is extraordinarily reactive, and generally, there are no crucibles capable of holding liquid iron oxides except Fe and graphite ones. There is no need to use a crucible in the gas conveyed system since the fine molten iron oxide droplets are dispersed by suspension in the flowing gas. (2) The mass transfer resistance becomes negligibly small due to the very small size of iron oxide droplets, allowing the true chemical reaction rates to be directly measured.

Tsukihashi *et al.* [2] investigated the reduction kinetics of molten iron oxide with CO gas using a gas conveyed reacting system. Their experimental design was shown in Fig. 4-5 (Chapter 3).

In their experiments, the dispersed fine oxide particles (with a mean diameter of about 25 µm) were quickly melted during falling, and they became spherical, and it was found that the reduced iron was surrounded by liquid FeO, as shown in Fig. 8-3. Therefore, the reduction reaction was assumed to have occurred at the molten iron oxide surface through the reduction process, and the formed Fe did not disturb the CO reduction at the surface.

The apparent reduction rates were almost independent of the initial ratio of Fe^{3+}/Fe^{2+} in the iron oxide powder. They also found that small Fe particles were formed during the early stage of the reduction. These results mean that the reduction without Fe formation was rapidly finished in the early stage of the reduction, and most of the part of the reduction process was carried out with the Fe saturated condition. Thus, neglecting the early stage of the reduction process and presenting molten iron oxide as 'FeO'(l), the overall reduction reaction in their experiments is given by

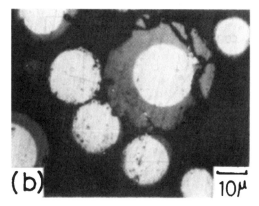

Fig. 8-3 The partially reduced quenched molten iron oxide particles in the gas conveyed reacting system. The Fe spherical particle (white part) is surrounded by a molten iron oxide layer (dark region) [2].

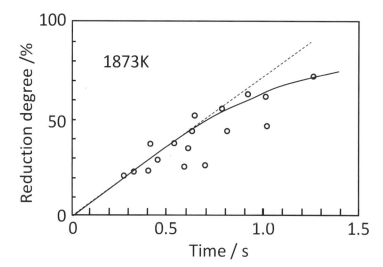

Fig. 8-4 The reduction degree of molten iron oxide with CO gas as a function of time at 1873 K [2]

$$\text{'FeO'(l)} + CO \rightarrow Fe + CO_2. \qquad\qquad ----- (8\text{-}5)$$

The reduction degree of the molten iron oxide with CO gas as a function of time at 1873 K is shown in Fig. 8-4.

The diameter of the molten iron oxide droplet gradually decreased due to the formation of the Fe core since the density of molten Fe is higher than molten iron oxide. Tsukihasi *et al.* evaluated the reaction rate by considering the change in the diameter during the reduction process. For details, refer to the original paper [2]. The calculated overall reaction rate R_T using the gradient of the broken line in Fig. 8.4 is about 1.7×10^{-5} mol/cm²·s.

The mass transfer coefficient m in their experimental conditions was estimated to be about

2900 [cm/s] or 0.019 [mol/cm²s·atm] at 1873 K by using the Ranz-Marshall equation (see section 3.8). That is, m is so much larger than the measured apparent rate that the mass transfer resistance is negligible, or the effect of produced CO_2 at the interface on the overall rate is insignificant.

Thus, the overall reaction rate R_T [mol/cm²s] of the reaction (8-5) can be expressed by

$$R_T = k_T P_{CO} \qquad ----- (8\text{-}6)$$

where k_T [mol/cm²s·atm] is the overall reaction rate coefficient. k_T is given by

$$1/k_T = 1/k_C + 1/m \qquad ----- (8\text{-}7)$$

where k_c [mol/cm²s·atm] is the chemical reaction rate coefficient. Since $1/k_T$ (5.8 × 10⁴) >> 1/m (52.6), Then, $1/k_T \approx 1/k_c$ or $k_c \approx k_T$, the measured rate constant k_T of 1.7 × 10⁻⁵ mol/cm²s·atm may represent the chemical reaction rate k_C . The chemical reaction rate coefficient k_c of the forward reaction (8-5) under the Fe saturate condition is given to be about 1.7 × 10⁻⁵ mol/cm²·atm s at 1873 K and about 1.2 × 10⁻⁵ mol/cm²·s atm at 1723 K.

8.2.2 Reduction of molten iron oxide held in a Fe crucible with CO

Nagasaka *et al.* carried out the CO gas reduction of molten iron oxide held in a shallow Fe crucible by using the CO-Ar gas mixture [3]. The used experimental system was shown in Chapter 3. In their experimental conditions, the molten iron oxide was always saturated with Fe, and the Fe^{3+}/Fe^{2+} ratio in the molten iron oxide did not change during the reduction. They found that Fe did not cover the surface of the molten iron oxide up to the reduction degree of 10%.

The gas reduction rate of molten iron oxide was evaluated based on the weight loss at 1673 K using a thermal balance. The reduction degree of molten iron oxide with Ar- 10%CO and Ar- 3%CO gas mixtures as a function of time at 1673 K is shown in Fig. 8-5.

The reduction progresses with time until the reduction degree of about 10%, but it gradually decreases after that. They confirmed that the reduction rate was controlled by the chemical reaction at the interface with the gas flow rate above 2.5 *l*/min. In the range of 0.02 < pCO < 0.18 in the Ar-CO gas mixture, the reduction rates were found to be proportional to pCO as shown in Fig. 8-6.

Thus, the apparent rate was given by the following equation.

$$\text{Rate} = k_a \text{pCO} \qquad ----- (8\text{-}8)$$

where k_a is the apparent reaction rate coefficient of the reaction (8-5). From the gradient of the reduction curve of the linear part (up to 10%) shown in Fig. 8-5, the apparent reduction rate coefficient is calculated to be about 1.1 × 10⁻⁵ mol/cm²s·atm at 1673 K. The mass transfer coefficient m is calculated based on the mass transfer correlation proposed by Kikuchi *et al.* and is about 10 times larger than the apparent rate coefficient. Using the same method as that in the previous section, the overall reaction rate coefficient k_a of 1.1 × 10⁻⁵ mol/cm²s·atm can be regarded as the chemical reaction rate coefficient of the reaction (8-5) at 1673 K. The chemical

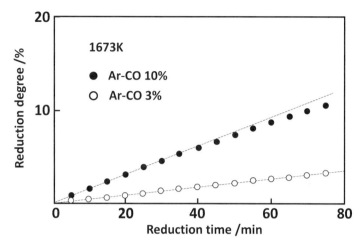

Fig. 8-5 The reduction degree of molten iron oxide with Ar- 10%CO and Ar- 3%CO gas mixtures as a function of time at 1673 K [3].

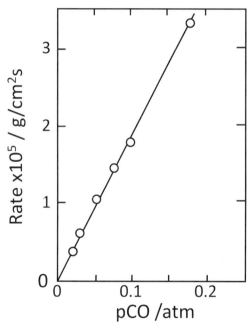

Fig. 8-6 The apparent reduction rate of molten iron oxide with CO in the range of 0.02 < pCO < 0.18 with the gas flow rate of Ar-CO gas mixture above 2.5 l/min. [3]

reaction rate coefficient at 1723 K is also given to be about 1.5×10^{-5} mol/cm^{2}s·atm. This value is reasonably close to that of Tsukihashi *et al.* [2].

8.3 Reduction of molten iron oxide without Fe formation

8.3.1 Isotope exchange reaction of CO$_2$ dissociation

In the reduction of molten iron oxide under the Fe saturated conditions, the Fe^{3+}/Fe^{2+} ratio in molten iron oxides is fixed. (*e.g.*, the ratio Fe^{3+}/Fe^{2+} with Fe saturated condition at 1673 K is about 0.084.) The Fe^{3+}/Fe^{2+} ratio in the molten iron oxides changes not only with pO$_2$ in the atmosphere but also by the addition of other oxides, such as CaO, Al$_2$O$_3$, and SiO$_2$. The effect of these oxides is discussed in the following chapters. Sasaki *et al.* [4] carried out the following isotope exchange reaction on molten iron oxides with varying pCO$_2$/pCO ratios to evaluate the effect of Fe^{3+}/Fe^{2+} on the oxidation or reduction rates on the molten iron oxide with CO$_2$-CO gas mixtures,

$$^{14}CO_2 + {}^{12}CO \rightarrow {}^{14}CO + {}^{12}CO_2 \qquad \text{----- (4-6)}$$

That is, CO$_2$-CO mixtures with CO$_2$ containing ^{14}C were passed over molten iron oxides, and the produced ^{14}CO was measured.

To avoid the contribution of the exchange reaction occurring at the refractory surface of the experimental system on the overall reaction, they developed the unique experimental system shown in Fig. 8-7.

Briefly, a small sample (about 0.5 g) of iron oxide was held as a molten film in a shallow well in an inductively heated Pt/Rh alloy susceptor held by a one-end closed silica sheath tube. The sample temperature was measured by a noble metal thermocouple inserted in the sheath tube. An alumina containment tube was attached to the well. The gas sealing between the well and containment tube was maintained by the surface tension of molten iron oxide. This sealing prevented the exposure of the Pt/Rh surface to the reaction gases, which were passed to the surface of the melt through a coaxially held alumina tube. The amounts of ^{14}CO before and

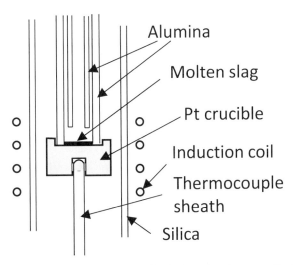

Fig. 8-7 Experimental arrangement for the isotope exchange reaction [4].

after the exchange reaction were evaluated using the Geiger-Müller tube method.

The overall isotope exchange reaction of (4-6) on molten iron oxide surface occurs by the consecutive steps:

$$^{14}CO_2 \rightarrow {}^{14}CO + O^* \qquad\qquad ----- (8\text{-}9)$$
$$^{12}CO + O^* \rightarrow {}^{12}CO_2 \qquad\qquad ----- (8\text{-}10)$$

where O^* stands for some of the oxygen at the surface involving in the reactions (8-9) and (8-10). It is not known whether the oxygen provided from the CO_2 stays as an adsorbed oxygen at the surface for a while or whether it is quickly incorporated in the liquid iron oxide. Thus, the oxygen involved in the reaction of (8-9) and (8-10) is tentatively presented by O^*. Because there is no way to evaluate the O^* behavior during the reaction, the effect of oxygen in the isotope exchange reaction on the iron oxide surface is not explicitly expressed in the rate equation but is evaluated by incorporating the oxygen effects into the rate coefficient, or $k_a = k_a(pO_2, T)$.

The reaction rate of the reaction (8-9) is presented by

$$dn(^{14}CO_2)/dt = k_a p^{14}CO_2 \qquad\qquad ----- (8\text{-}11)$$

where k_a is an apparent rate constant of the reaction (8-9). k_a is given by the following expression from the measurement of the produced amount of ^{14}CO in the experiments [5, 6]. The eq. (8-12) **does not depend on the detailed reaction mechanism.**

$$k_a = \frac{V}{ART} \frac{1}{1+B} \ln\left[\frac{1}{1-p^{14}CO/(p^{14}CO)_{eq}}\right] \qquad ----- (8\text{-}12)$$

where V is the total volume flow rate of the gas mixture exposed to an area A of the melt, and B is the CO_2/CO ratio. $p^{14}CO$ is the partial pressures of tagged CO after the reaction, and $(p^{14}CO)_{eq}$ is that for the case when complete isotope equilibrium is achieved, respectively. It is possible to determine all the values required in these equations of (8-12) by taking experimental measurements.

The calculated values for the apparent rate constants for liquid iron oxide at 1693 K and 1773 K as a function of pCO_2/pCO are presented in Fig. 8-8, where the unit for the rate constant k_a is $mol\ cm^{-2}s^{-2}atm^{-1}$. The slope of the line through the date is about -1.0 for both temperatures. Accordingly, the apparent rate constant k_a is expressed by

$$k_a = k_0(pCO_2/pCO)^{-1} \qquad\qquad ----- (8\text{-}13)$$

where k_0 is a temperature-dependent constant. That is, the apparent rate constant k_a is not constant but strongly depends on the pCO_2/pCO ratio. The dependency of (8-13) means that the oxygen effect is incorporated into the rate coefficient, or $k_a = k_a(pCO_2/pCO, T) = (pO_2, T)$.

The apparent rate constants as a function of temperature for a fixed CO_2/CO ratio of 1 [4] are presented in Fig. 8-9.

From the gradient in the Fig. 8.9, the k_0 in the equation (8-13) is presented by the equation since $k_a = k_0$ with CO_2/CO ratio of 1,

$$\log k_0 = -6900/T - 0.88 \qquad\qquad ----- (8\text{-}14)$$

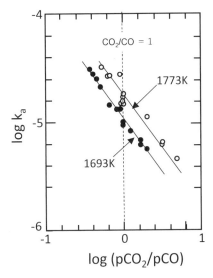

Fig. 8-8 Apparent rate constants (mol cm^{-2} s^{-1} atm^{-1}) for the dissociation of CO_2 on liquid iron oxide at 1693 K and 1773 K as a function of the equilibrium CO_2/CO ratio [4].

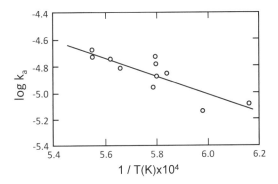

Fig. 8-9 Temperature dependence of the apparent rate constant for the dissociation of CO_2 on liquid iron oxide at a CO_2/CO ratio of 1 [4].

The CO_2 dissociation reaction on the molten Fe surface does not occur in the area covered with adsorbed oxygen. However, it occurs on the molten iron oxide surface, even though the surface of the molten iron oxide is fully covered with oxygen. This means that the surface structure of the adsorbed oxygen is different from that of oxygen at the surface of the iron oxide.

The reaction of CO with molten iron oxide

The following isotope exchange reaction on molten iron oxides

$$^{14}CO_2 + {}^{12}CO \rightarrow {}^{14}CO + {}^{12}CO_2 \qquad\qquad ----- (4\text{-}6)$$

occur by the consecutive steps:

$$^{14}CO_2 \rightarrow {}^{14}CO + O^* \qquad\qquad \text{----- (8-9)}$$

$$^{12}CO + O^* \rightarrow {}^{12}CO_2 \qquad\qquad \text{----- (8-10)}$$

Since the rate of the unlabelled CO_2 dissociation reaction (8-9)' is practically the same as that of $^{14}CO_2$.

$$^{12}CO_2 \rightarrow {}^{12}CO + O^* \qquad\qquad \text{----- (8-9)'}$$

Under the equilibrium condition, the reaction rates of (8-9)' and (8-10) are equal. As already mentioned, the role of O^* is incorporated into the reaction constants. Thus,

$$k_1 pCO_2 = k_2 pCO \qquad\qquad \text{----- (8-15)}$$

Combined with (8-13) and (8-15), k_2 is presented by

$$k_2 = k_0 \qquad\qquad \text{----- (8-16)}$$

That is, the reduction rate constant k_2 of the reaction (8-9)' is presented by k_0 expressed by (8-14). Since k_0 depends on the only temperature, k_2 does not change with pCO_2/pCO. The CO reduction of liquid iron oxide is not affected by the surface oxygen properties, which is different from the CO_2 dissociation reaction on the liquid iron oxide surface.

In the molten iron oxide reduction with CO by Tsukihashi et al. [2] and Nagasaka et al. [3], it was found that the apparent rate was expressed by $k_{ap} pCO$. Their results are shown in Fig. 8-10. The rate constant established using isotope exchange reaction described by eq. (8-16) is also superimposed. All of these results show good agreement.

Tsukihasi et al. [2] and Nagasaka et al. [3] carried out molten iron oxide reduction using 100% CO gas. That is, no oxygen was supplied from CO_2 to the molten iron oxide surface during in the reduction process. This is different from the isotope exchange reaction, where oxygen was supplied. As shown in Fig. 8.10, the reduction rate constants among these three experiments show a good agreement. This means that incorporating adsorbed oxygen into the liquid oxide structure provided from CO_2 is sufficiently fast that it is at virtual equilibrium. In

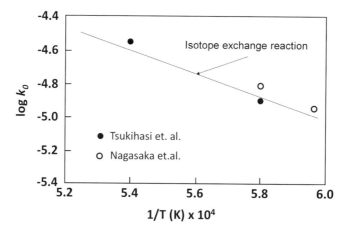

Fig. 8-10 Comparison of the apparent rate constants for the CO reduction of molten iron oxide by isotope exchange reaction and these of Tsukihasi et al. [2] and Nagasaka et al. [3].

other words, the CO_2 dissociation step involves the rapid incorporation of the supplied oxygen into the oxide. That is, there are no differences between the adsorbed oxygen (O*) and the lattice oxygen at the surface, at least not chemically.

The overall reaction of Fe-saturated molten iron oxide reduction with CO is presented by

$$\text{`FeO'(l)} + CO \rightarrow Fe + CO_2. \qquad\qquad ----- (8\text{-}5)$$

This reaction can be separated into the following two steps:

$$CO + O^{2-} \rightarrow CO_2 + 2e^- \qquad\qquad ----- (8\text{-}1)$$

$$Fe^{2+} + 2e^- \rightarrow Fe \qquad\qquad ----- (8\text{-}3)$$

The agreement among three different experimental results shown in Fig. 8.10 means that the reaction of (8-3) is faster than (8-1), and the reaction (8-1) is controlled by (8-5).

8.4   Effect of CaO on the molten iron oxide reduction

Sasaki *et. al.* [4], and Sun *et. al.* [7] measured the CO_2 dissociation rates on CaO-saturated liquid iron oxide using a $^{14}CO_2$-CO isotope exchange reaction with varying CO_2/CO ratios or pO_2. The simplified phase diagram of the Fe_2O_3-FeO-CaO system at 1723 K is shown in Fig. 8-11 [8]. The grey colored area is liquid phase region. The numeric values in the figure mean the oxygen partial pressure (atm).

The CaO content in the CaO-saturated liquid iron oxide system slightly decreases with the increase of pO_2. The slight CaO decrease with pO_2 is also observed at other temperatures. The effect of the change in the CaO content on the CO_2 dissociation rates is negligible since the change in the CaO content is small compared with the Fe^{2+}/Fe^{3+} ratio.

Experiments were carried out with the CaO containment tubes instead of Al_2O_3 ones. A typical setup is shown in Fig. 8-12.

The evaluated rate constants (mol cm⁻²s⁻¹atm⁻¹) for CaO saturated liquid iron oxide at 1823 K, and 1773 K as a function of pCO_2/pCO are presented in Fig. 8-13. The slope of the line

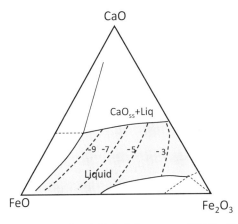

Fig. 8-11   The phase diagram of Fe_2O_3-FeO-CaO system at 1723 K with isobar of pO_2 [8].

134

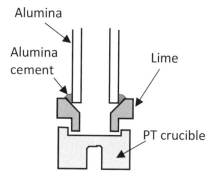

Fig. 8-12 The arrangement of the container tube used for the isotope exchange studies of CaO saturated liquid iron oxides [4, 7].

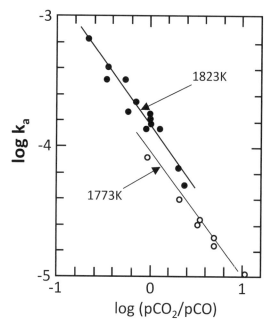

Fig. 8-13 Apparent rate constants (mol cm^{-2} s^{-1} atm^{-1}) for the dissociation of CO_2 on liquid iron oxide at 1773 K and 1823 K as a function of the equilibrium CO_2/CO ratio [4].

is about -1.0 for both temperatures. Accordingly, the apparent rate constant k_a is given by

$$k_a = k_0(pCO_2/pCO)^{-1} = k_0(a(O))^{-1} \qquad \text{----- (8-17)}$$

where k_0 is a temperature-dependent constant and is given by the following equation:

$$\log k_0 = -1650/T - 3.01 \qquad \text{----- (8-18)}$$

The rate constants of the CO_2 dissociation reaction on the molten iron oxide and CaO-saturated iron oxide as a function of pCO_2/pCO at 1773 K are shown in Fig. 8-14. With the same CO_2/CO condition, the CaO saturated molten calcium ferrite rate is about 1 order faster than molten iron oxide. When considering the CO_2 dissociation rate on molten iron oxide-containing CaO, the rate is expected to decrease with the increase of CaO content since iron oxide content at the surface in calcium ferrite melts is smaller than that in iron oxide without CaO. However, in reality, the rates for calcium ferrite melts are much more significant than iron oxide, as shown in Fig. 8-14. This means that the rates do not relate to the FeO content or FeO activity in the melt.

8.5   Effect of SiO_2 on the molten iron oxide reduction

El-Rahaiby et al. measured the CO_2 dissociation rates on silica-saturated iron silicates using a $^{14}CO_2$-CO isotope exchange reaction with varying CO_2/CO ratios [9]. The measured apparent rate constants in units of mol/(cm^2 s atm) with varying CO_2/CO ratios at 1513 and 1673 K are shown in Fig. 8.15.

Within experimental scatter, there is a linear dependency with the represented slope of -1. Thus, as in the case of liquid iron oxide and liquid CaO-saturated iron oxide, the apparent rate constant is presented by

$$k_a = k_0 (pCO_2/pCO) \qquad \text{----- (8-19)}$$

where k_0 is a temperature-dependent constant. Based on the results at different temperatures, the apparent reaction rate is presented by

$$\log k_a = -11100/T + 1.22 \qquad \text{----- (8-20)}$$

k_0 can be easily obtained by substituting $pCO_2/pCO = 1$ into (8-19).

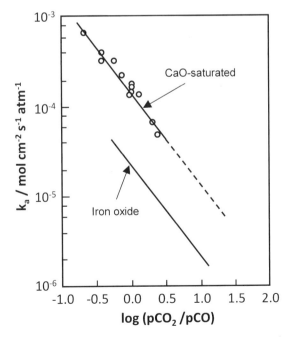

Fig. 8-14   Comparison of the apparent rate constant for the isotope exchange reaction on CaO saturated liquid calcium ferrites and liquid iron oxide.

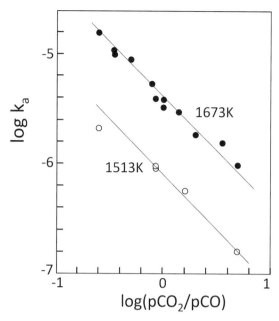

Fig. 8-15 Apparent rate constants for the CO_2 dissociation on SiO_2-saturated liquid iron oxide as a function of the CO_2/CO ratio.

8.6 Electron involvement in the reduction of molten iron oxides

The following equation shows the apparent reaction rate constants of molten iron oxide, CaO-saturated liquid iron oxide, and SiO_2 saturated iron oxide.

$$k_a = k_0 (pCO_2/pCO)^{-1} \qquad \text{----- (8-21)}$$

The pCO_2/pCO ratio or pO_2 will fix the Fe^{3+}/Fe^{2+} ratio in molten iron oxides, CaO-saturated liquid iron oxide, and SiO_2 saturated iron oxide. There are essentially linear relationships between the pCO_2/pCO ratio and the Fe^{3+}/Fe^{2+} ratio for these systems.

$$(Fe^{3+}/Fe^{2+})^2 = 0.027(pCO_2/pCO) \qquad \text{----- (8-4)}$$

for liquid iron oxide with the temperature-independent expression [1].

For CaO-saturated liquid calcium ferrites at 1773 K [10, 11],

$$(Fe^{3+}/Fe^{2+})^2 = 0.62(pCO_2/pCO). \qquad \text{----- (8-22)}$$

For SiO_2-saturated iron oxide [12]

$$(Fe^{3+}/Fe^{2+})^2 = 0.00133(pCO_2/pCO) \qquad \text{----- (8-23)}$$

Thus, the apparent reaction rate constants for molten iron oxide and CaO-saturated calcium ferrite melts, and SiO_2 saturated liquid iron oxide are proportional to $(Fe^{3+}/Fe^{2+})^{-2}$ and are shown in Fig. 8-16. 'FeO' refers to liquid iron oxide.

The inverse dependence of the apparent rate constant on the $(Fe^{3+}/Fe^{2+})^{-2}$ suggests that the transfer of two electrons is involved in the reaction in these liquid iron oxide systems. The oxygen in molten iron oxide exists as O^{2-}. The adsorbed oxygen O(ad) generated by the disso-

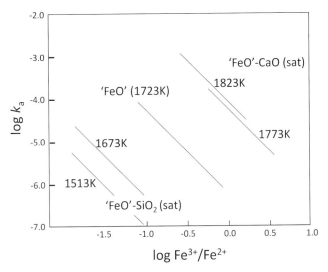

Fig. 8-16   The apparent rate constants for the isotope exchange reaction in liquid iron oxide, CaO-saturated liquid iron oxide, and SiO_2-saturated liquid iron oxide as a function of Fe^{3+}/Fe^{2+} ratio. 'FeO' refers to liquid iron oxide.

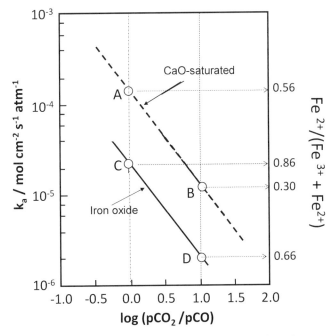

Fig. 8-17   The apparent rate constant for the isotope exchange reaction in CaO-saturated liquid iron oxide and liquid iron oxide with Fe^{2+} ion fraction at $pCO_2/pCO = 1$ and 10.

138

ciation reaction of CO_2 needs to obtain two electrons from somewhere and become O^{2-} for the incorporation into molten iron oxygen. As shown in Fig. 8-16, the CO_2 dissociation rates decrease with Fe^{3+}/Fe^2 in these three liquid oxide systems.

As shown in Fig. 8-17, the Fe^{2+} fractions ($= Fe^{2+}/(Fe^{2+} + Fe^{3+})$) at points A and B in CaO-saturated liquid iron oxide are 0.56 and 0.30, respectively. At points C and D in liquid iron oxide, the Fe^{2+} fractions are 0.86 and 0. 66, respectively. That is, the CO_2 dissociation rate increases with the increase of Fe^{2+} fraction in each melt. While comparing the Fe^{2+} fractions at points A and C, or B and D, the Fe^{2+} fractions in liquid iron oxide are larger values than in CaO-saturated liquid iron oxide. However, the rate constant of liquid iron oxide at pCO_2/pCO = 1, or 10, is about 1 order smaller. In other words, smaller Fe^{2+} fractions are associated with a faster dissociation rate than in different iron oxide systems. This apparent conflict of the role of Fe^{2+} strongly suggests that there are still unknown factors that affect the CO_2 dissociation reaction on the surface of molten iron oxides.

Further reading

L. von Bogdandy and H.-J. Engel The reduction of iron ore, Spring-Verlag Berlin, 1971.

E. T. Turkdogan: Physical Chemistry of High Temperature Technology, Academic Press, New York, 1980.

A. Ozaki: Isotope Studies of Heterogeneous Catalysis, Academic Press New York 1977.

References

[1] Darken Gurry: *J. Am. Chem. Soc.*, 1946, vol. 68. 798-816.

[2] F. Tsukihashi, K. Kato, K. Otsuka and T. Soma: *Trans. ISIJ*, 22 (1982), 688-95.

[3] T. Nagasaka, Y. Iguchi and S. Ban-ya: *Tetsu to Hagane*, 71 (1985), 204-211.

[4] Y. Sasaki, S. Hara, D. R. Gaskell and G. R. Belton: *Metal. Trans. B*, 15B, (1984), 563-571.

[5] A. Ozaki: Isotope Studies of Heterogeneous Catalysis, Kodansha Ltd., Tokyo and Academic Press Inc., New York, NY, 1977, 25-27.

[6] A. W. Cramb, W. R. Graham, and G. R. Belton: *Metal. Trans. B*, 9B (1978), 623-629.

[7] S. Sun, Y. Sasaki and G. R. Belton: *Met. Trans. B*, 19B, (1988), 959-965.

[8] M. Timucin and A. E. Morris: *Metall. Trans.*, 1 (1970) 3193-3201.

[9] S. K. El-Rahaiby, Y. Sasaki, D. R. Gaskell and G. R. Belton: *Met. Trans. B*, 17B, (1986), 307-316.

[10] Y. Takeda, S. Nakazawa, and A. Yazawa: *Can, Met., Quart.*, 19, 1980, 297-305.

[11] H. Larson and J. Chipman: *Trans. AIME*, 1953 vol. 197, 1089-96.

[12] E. J. Michel and R. Schumann: *Trans AIME*, 194 (1952) 723-29.

Index